Successful Program Management

Complexity Theory, Communication, and Leadership

Best Practices and Advances in Program Management Series

Series Editor
Ginger Levin

Successful Program Management

Complexity Theory, Communication, and Leadership

Wanda Curlee

Robert Lee Gordon

CRC Press
Taylor & Francis Group
Boca Raton London New York

CRC Press is an imprint of the
Taylor & Francis Group, an **informa** business
AN AUERBACH BOOK

CRC Press
Taylor & Francis Group
6000 Broken Sound Parkway NW, Suite 300
Boca Raton, FL 33487-2742

© 2014 by Taylor & Francis Group, LLC
CRC Press is an imprint of Taylor & Francis Group, an Informa business

No claim to original U.S. Government works

ISBN 13: 978-1-4665-6879-2 (hbk)

Library of Congress Cataloging-in-Publication Data

Curlee, Wanda.
    Successful program management : complexity theory, communication, and leadership / Wanda Curlee, Robert Lee Gordon.
        pages cm. -- (Best practices and advances in program management series)
    Includes bibliographical references and index.
    ISBN 978-1-4665-6879-2 (hardcover : alk. paper) 1. Project management. I. Gordon, Robert Lee. II. Title.

HD69.P75C874 2014
658.4'04--dc23   2013030837

Visit the Taylor & Francis Web site at
http://www.taylorandfrancis.com

and the CRC Press Web site at
http://www.crcpress.com

Contents

Introduction

Several years ago the Project Management Institute (PMI) took the bold step of defining a program. PMI's *The Program Management Standard* (PMI, 2013b) broadly defines programs as a selection of projects, subprograms, or activities to achieve a strategic goal with benefit realization at a lower cost. This definition is clearly the most effective definition for a program in this context because it does not limit a program to a single isolated build or to a set of related projects. Other definitions are too restrictive because they are narrow in scope or relationship. Although this conceptual direction of a program might be controversial, a few simple examples explain that this definition is not only effective, it is essential in defining a program.

Linear thinkers often view a program as a related collection of projects that are tightly or loosely aligned in a manner that adds value to an organization. In this regard, many related projects are brought together until an inclusive umbrella covers them. As an example of this kind of linear thinking, one would cluster the build program for two different Navy vessels, say a support vessel such as an oiler and a fighting vessel such as an aircraft carrier. On the face, one could conceptualize different synergies that might conclude in a lower cost when combining the two projects. Traditional thinking could combine these two projects into one program, where the combined projects might result in lower costs as one shipyard might be willing to accept a smaller profit on each vessel in order to secure two large projects. This type of linear thinking clearly has its place in program management; however, it is often where program managers stop. Linear thinking does not allow for taking program management to the level PMI has envisioned.

Non-linear thinkers understand the intent of PMI and hence are willing to look beyond the obvious in order to achieve the strategic goal of lower costs. Non-linear thinkers will seek to combine seemingly unrelated projects in a manner that achieves lower costs. For example, one can combine an information technology (IT) project, the construction of a new fighter jet, and the remodeling of a new office building in a manner that can reduce costs for the purpose of a strategic goal.

One must look at the traditional areas where one might find a diversified contractor who might be able to combine these procurements into a traditional program, but in addition, one must look at certain areas that make sense together to expand the use of an encompassing program. If these projects are in a similar geographic area, then one might seek to combine all the recycling and re-using of all the related materials. Although the spend for the new fighter jet would likely be the most costly of the three projects, if recycling and re-using materials, one might find the most opportunity with the remodeling project. By combining these projects, one could create a program that achieves a green goal for the organization while achieving greater efficiencies with regard to recycling and re-using and hence drive down the costs for all of the projects.

Another example of a program that might benefit all three of these component projects is leveraging a single commodity item or group of items that might benefit all the component projects. One might not see any related materials, but one must look beyond. One commodity that might touch upon these programs is cable. There will be IT and other electrical cable in the fighter jet, there will be IT and other electrical cable used in the office remodel, and there will likely be cable used in the IT project. By combining all of these volumes, costs could be reduced for all the projects. An example of a single item that would be used for these projects would be personal computers. Since these component projects will require the purchasing of software and hardware, a program manager should seek to combine the projects into a program that would drive down the costs of software and hardware for all these projects.

Thus, by deploying program managers who can manage seemingly unrelated diverse projects, an organization can better achieve its goals while reducing costs. One might feel that these additional program managers increase costs, but since their focus is aimed at cost reduction and organizational improvement, the labor costs will be offset by the reductions. Furthermore, by deploying these program managers strategically, they can help achieve organizational goals that might be overlooked by smaller projects with fewer resources. There are no limits to how creatively a program manager can be deployed within an organization, and a program should not be limited to apparently related projects.

SUMMARY OF THE BOOK

This book is a study of program management, but it focuses on communication via the essentials of leadership and how understanding complexity theory aides the program manager. Understanding the nuances of complexity theory and its sub-theories assists the program manager by helping him or her understand where to focus strategic energies and arm the program team with the necessary skills, tools, and techniques to succeed.

Academics and practitioners will both find information of interest because the needs of each of these groups will be addressed. The academics will note that the information is supported by peer review research and each section has a case study, section quiz, and discussion questions. This offers not only a means to learn but a means to apply this practically in a classroom setting. Practitioners will discover the book has numerous tools, templates, and techniques to help the seasoned program manager as well as the program manager who is the leader of a program for the first time. These tools offer approaches and ideas that can be deployed by program managers in the field with both successful and less than successful programs.

Complexity is a part of a program and hence complexity theory. To understand complexity theory from the business perspective and not on a purely scientific level, the book leads the reader through complexity leadership and communication.

1

Introduction

PHOTOGRAPH 1.0
Complexity can be seen in the natural world with the interaction of the sky, the clouds, the trees, and the field of mustard.

HISTORY OF COMPLEXITY THEORY

Humans like to think of the world as nice and orderly. Many people may say their lives are quite boring because their lives are so predictable. Airline pilots have been known to state that flying is hours of boredom interjected with a few seconds of sheer panic. Whether we like it or not, life presents interruptions to our daily routine. These may be minor or cataclysmic, but nonetheless they are interruptions.

The minor interruptions we have learned to deal with in our everyday lives. These may range from the dog needing to go out when we are on the phone to an impromptu conference call held by our supervisor while we are in the middle of writing an important memo. These are annoyances that most of us have learned to deal with, but they are part of complexity theory, as we will learn a little later.

The major interruptions can be life changing and/or catastrophic to a program and must be dealt with, normally not by the program manager. These situations are discussed in more detail in later chapters. The September 11, 2001, terrorist attack in the United States and subsequent response by the Federal Aviation Administration (FAA); the Deepwater Horizon disaster in the Gulf of Mexico and the command of the fleet to clean the ensuing oil disaster; and Japan's natural disasters which ultimately led to the Fukushima Daiichi nuclear power plant disaster are examples of how linear thinking and non-linear thinking can be used effectively in complex situations.

Complexity theory applied to a program may be defined as those areas that are on the *edge of chaos*. The size and complicated nature of the program drive the number of edges of chaos. Since the program manager is not available to deal with each edge of chaos that flips into chaos, the program manager must train his/her team leads and the team to deal with the chaos. This does not mean the program manager abdicates his/her authority but rather that he/she instills trust in the team through adequate communication.

Complexity theory has its roots in chaos theory, which was proposed by Edward Lorenz, a meteorologist (Wheatley, 1999). Lorenz offered an explanation of the effects of chaos theory with the theory of the butterfly effect. The butterfly effect can be explained simply: when a bird flaps its wings in Florida this creates a minute disturbance in the atmosphere, which unto itself does not appear to be significant but combined with a multitude of other factors may turn into a hurricane in some other region. Weather cannot be precisely understood, because one would have to mathematically take into account every possible atmospheric disturbance. Since some atmospheric disturbances are very small, such as the example of the bird flapping its wings, this makes predicting the exact path of a hurricane difficult. Even modern techniques of weather prediction can only approximate the movement of significant weather. When weather events could not be tracked or calculated, they were originally believed to be random, and many scientists have attempted to discount these smaller forces as irrelevant due to their relatively small impact. Lorenz found that the atmosphere never reached a state of equilibrium; it is always in a state of chaos and so every factor has an impact, even if not perfectly understood. When measuring so-called randomness, the diagrammed results were in the shape of butterfly wings or owl eyes. Hence the name *butterfly effect*.

As complexity theory evolved and was applied to social sciences, experts point to how the impact of a single leader can make a difference. Lincoln had not held an elected office until he secured the presidency, and Grant was fairly unsuccessful in his wife's family business prior to the Civil War. How different today would be had both of these great men not risen to their true potential. No one could have predicted that both would have had such a pivotal impact on history prior to the Civil War.

History has shown that the difference between victory and defeat can rest upon the shoulders of a single individual. A single, bold leader can make a difference in a battle and hence in an entire war. A single leader has the potential to motivate thousands in a manner that allows for the completion of complex tasks. In addition to the potential of a single great leader, small contributions can build to create something larger than their individual parts. A leader who can motivate people is rarely the person who does it all, and so one must understand that the leader's impact is based upon the seemingly small changes that touch others. A leader can offer small praises that matter, which help move a program forward. The more that a program manager, particularly one who operates virtually, can harness this kind of organization that passes this feedback forward, the more effective the organization will become.

Complexity theory acknowledges that humans when working together are a complex open system. It differs from the traditional open systems theory in that complexity theory acknowledges there are elements in the system that cannot be explained. Human interactions are complex and not always consistent, and unlike technology that is designed to work only in a certain manner or else not work at all, humans are able to work at totally different levels for totally different reasons. Because of this, certain elements must be considered as a normal part of randomness (Byrne, 1998). Humans like to believe that they understand all of these elements and that they can predict risk, but people can only estimate risk. Human beings like to break down a system to the smallest part in order to explain the whole system. This kind of thinking is based upon understanding complex systems, such as the universe, but one must understand the discrete systems rather than the whole.

Although seductive, this approach would fail in many cases because one cannot examine how a single ant works and then make correct assumptions about the entire colony. Studying a single organ such as the human heart does not explain the interrelationship of the glands, brain, heart, blood, and so forth. What complexity theory offers is a methodology to harness groups in a manner that allows the program manager to increase the team's effectiveness by allowing a certain degree of individuality to move a program forward (Hass, 2009). Oftentimes allowing the random walk of the determined individual allows that individual creative latitude to become successful. An effective team can be more effective than a single individual, because a team can generate more good ideas than a single individual.

Seasoned program managers realize that all parts of the program cannot be controlled and that they would not want to have full control of the program. They realize that creativity occurs on the fringes of complexity or chaos. Linear thinkers can only consider evolutionary thoughts where one process connects to another sequentially with the potential of incremental improvement, while complexity-based thinkers can create revolutions and introduce new systems with the potential for enormous improvements.

Complexity goes beyond chaos theory and has been applied in social sciences and business. The butterfly effect can be applied and used effectively in large-scale virtual programs. From a program management perspective, a successful program is when all the available forces work in the same direction. Just as the flapping wings of a butterfly in Japan can be a contributing force to the creation of a hurricane in Florida, even a small impact can have a great impact when magnified over time and distance.

COMPLEXITY THEORY DEMYSTIFIED

To demystify complexity theory, one must examine different ideas that are non-linear and that may already be familiar but were not recognized as complexity theory. Five complexity theory concepts will be discussed in order to make complexity theory more approachable. As we have stated, complexity theory is a non-linear approach to program management. A program manager will likely find this not to be an intuitive form of leading a program; however, there are advantages to a non-linear approach. For example, the majority of the program manager's interaction with others involved with the program will be virtual; hence a program manager must learn to leverage both complexity and the virtual environment. The program manager needs to have the self-confidence to understand which areas of the complexity or chaos need focus and which areas the team should be allowed to resolve.

Below are five practical program applications of complexity theory that will help to demystify the application. It is likely that the experienced program manager has intuitively applied complexity theory to a program. Human beings are subject to open systems theory, and as such we adapt to complexity theory more than people realize.

Practical Application One

For the complex program, complexity theory suggests that costs (and risks) should be forecasted following each butterfly effect episode (program milestone). Strict cost forecasting should not be imposed, as it may limit the effectiveness of the process (Overman & Loraine, 1994).

The seasoned program manager understands that program failure is always related to limitations—of resources, of processes, of procedures, of management decisions, or of time. The program manager should forecast costs and risks after every milestone event related to the program. Furthermore, the program manager should not impose strict constraints upon any forecasting in order to make sure that the process is effective. A program manager might want to limit the process in order to save time, but ideas are born from other ideas. Limiting the imagination and possible ideas in this process can stifle the process. The nuclear issues in Japan occurred due to a limitation of imagination in the process. Designers created safeguards for seismic activity and for tsunamis but never considered how both might interact, despite both being identified as possible threats to the nuclear reactors.

Practical Tool One

One possible strategy is to forecast backwards. Imagine the completed program and consider each benefit and review each component. It is often best to do this by reviewing the timeline backwards and evaluating every piece of the puzzle. Make sure to consider contingencies at difficult milestones or connected events. Most importantly, consider the people issues. Imagine what would happen to the program if a key program manager or component project lead were to leave the program. Consider the impact of the loss of key people throughout the life of the program.

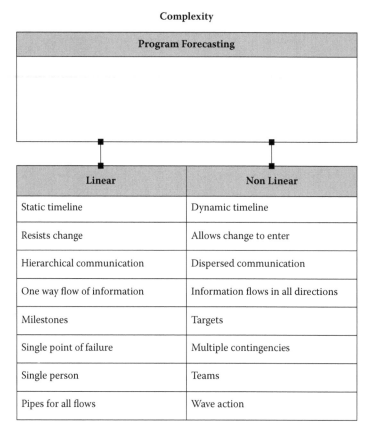

Complexity

Program Forecasting	
Linear	**Non Linear**
Static timeline	Dynamic timeline
Resists change	Allows change to enter
Hierarchical communication	Dispersed communication
One way flow of information	Information flows in all directions
Milestones	Targets
Single point of failure	Multiple contingencies
Single person	Teams
Pipes for all flows	Wave action

FIGURE 1.1
Complexity forecasting.

Figure 1.1 offers ideas to arrange for complex forecasting that take into account both linear thinking and non-linear thinking. One should put the ultimate goal at the top and then utilize linear and non-linear approaches to achieve the goal. Different ideas can then be noted to compare how the approaches might work separately or together. It is recommended that more non-linear approaches be used, but at the very least one should strike a balance.

Practical Application Two

Most program managers understand that all systems are connected, and by understanding these interconnections, a new understanding can be achieved regarding the program. Too often, program managers see each component project as a discrete entity, and this view does not explain the bigger picture. Understanding the role of the quarterback in football does not explain how football is played.

Program managers like to believe they have the program under constant control. Unfortunately, life (and Murphy's Law) gets in the way and unexpected issues are going to arise. All the risk planning, issue resolution, and management will not prevent things from happening. Program managers should understand that small or catastrophic events outside of the program may or may not affect the program. The same is true with events internal to the program.

As an example, Japan has earthquakes on a regularly occurring basis. Modern construction is designed to account for moderate earthquakes. As stated in the previous tip, Japan's government leaders and TEPCO leaders (those in charge of the nuclear reactors) neglected to account for or understand, or even ignored the interconnection of earthquakes and tsunamis. To address these kinds of related issues, a program manager may decide to consider that issues are not isolated. They may affect implementation, training, cost, schedule, among other items. Team leads must be taught and given the opportunity to practice to understand how the program interacts with the environment and other parts of the program. The program manager should schedule brainstorming sessions to help component and team leads understand these related issues and concepts.

Application Tool Two

Map out the tasks of the program using the Program Web Tool (Figure 1.2). The tool will force the program team to map the tasks in a non-linear fashion. Evaluate each task as it interacts in the program and the adjacent task. Considering the tasks in a linear fashion will lead to greater appreciation of the individual tasks as well as to an understanding of how they support the program goal. The program manager must review even the tasks that are not on the critical path. If a program starts having problems, an unrelated task might in a few days or a week be on the critical path.

Practical Application Three

Many organizations spend a lot of energy affixing blame to individuals; however, it is more productive to offer solutions than blame. In the end, problems that arise are rarely the fault of a single individual, and so it

FIGURE 1.2
Program web.

becomes difficult to find the one person to blame. Instead, more time should be spent with the solution, as that will ultimately be more responsible than one person.

Blame is easy; solutions are difficult. As a program manager you have the obligation and maybe even the moral/ethical duty to stay above gossip. It is easy to fall into the trap, and each of us does that at times. Catch yourself and stop. Bring your team together for a brainstorming session. Bring different teams together from their areas, but also remember to have cross-functional teams as well.

Cross-functional teams are also known to develop highly innovative solutions. Make them a standard part of the team process. The team may push back at first, but be persistent and patient.

Application Tool Three

When people start looking for someone to blame, remind them that the blame game never saved a program. There is a saying, that "when you point a finger at someone else, there are four fingers pointing back at you."

Practical Application Four

When considering the power of complexity theory, keep in mind the metaphor of a swarm of ants. Does the organization operate with a vision and a mission statement that motivate each individual toward an appropriate goal? The program manager is responsible for setting the goals, and the team leaders are responsible for setting the course for the program. Each might set the course in a different manner, and some may need some course adjustment, but unless a team leader is not heading toward the goal, the program manager should just monitor through dashboards and one-on-ones. Going down the chain, the individuals on the team should also be given latitude as long as the team is achieving the goal of the organization. Watch an anthill. The ants work in harmony with minimal communication and no communication from leadership. Interrupt the path to the anthill. The ants quickly start to implement a plan of attack (individually), again with minimal communication, and no communication with leadership to achieve the status quo.

Application Tool Four

Do a quick poll of the stakeholders of a program by asking each person what the goal of the program is and what each person wants to receive from the program at its completion. The results will offer keen insight to what people think about the program and what they think the program will do for their career.

Practical Application Five

People are complex, and their decision-making process is often not linear. People can be motivated by many different factors that may not be related to the program. Consider the replacement costs of top employees and compare that to the cost of addressing conflict. It is far more economical to keep a good team together than to try to build a new one whenever morale drops. Some organizations realize the cost it takes to replace good talent. Not only are there tangible costs of recruitment and training, but there is always a period of time for a replacement to come up to speed. This lost productivity is never regained, but the hope is that over time, the person will be productive enough in the future to regain the previous efficiencies. In general, it is cheaper and easier to address the human issues than to have to continually replace people over time.

Practical Tool Five

Consider what the loss of the three top people in a program would cost the program in a matter of time, productivity, and direct and indirect costs. Come up with a value and how much delay could be expected. Determine a recovery strategy for each person, and keep this information available in the event the program loses key people.

In conclusion, there are many complexity theory ideas that might already be deployed by an organization. One must be open to what is possible and consider how these possibilities might already apply or consider how they can be applied to a complex program. There is much to garner from complexity theory, so one must reflect upon these ideas in order to understand how they already apply or how they can apply to different programs in the future.

OVERVIEW OF PROGRAM MANAGEMENT

Program and *project* were interchangeable terms for most industries and companies until the Project Management Institute (PMI) published the first *Program Management Standard* in 2003. The standard took a definite stance that a project and a program had different aims. A project was tactical in nature, while a program was strategic.

Still today, some industries and companies define programs as large, complex projects. Another definition for a program may be a set of projects that are similar in nature and may or may not be led by a "program" manager. For example these similar projects may be installing cable televisions, Internet, and/or Internet phones on a daily basis.

PMI's Program Standard defines a program as "a group of related projects, subprograms, and program activities that are managed in a coordinated way to obtain benefits not available from managing them individually" (2013b, p. 166). A program manager is expected to monitor overall benefits by manipulating the component projects for the benefit of the program. This may delay a project or cause it to go over budget, but when the program manager reviews all the component projects, he or she most likely is accelerating another project to help the program achieve benefit realization.

The projects need to have interdependencies to be within a program. Without the interdependencies there might be some doubt as to whether the project merits being part of the program. The projects also do not need to begin or close at the same time. In fact, in most programs, projects are beginning and closing throughout the life cycle of the program. The program manager needs to understand how each project is interdependent because any schedule delays, schedule accelerations, budget overruns, or schedule under-runs will affect benefit realizations. The program manager must react, communicate, and keep the program on course while keeping morale and trust high.

The program manager needs to also understand that there are projects that are not part of the program that may affect the program. There will be other external factors affecting the program that will be discussed in other chapters of this book. Such items include communication skills, leadership, and risk mitigation.

Figure 1.3 provides a graphical representation of a typical mature project management organization. The top part is the program. The middle part of the project management organization is the project. The bottom part of the triangle is all the people. Some projects are part of programs, while others are independent projects. The next part and a bit smaller are the programs. As expected, there should be fewer programs than projects.

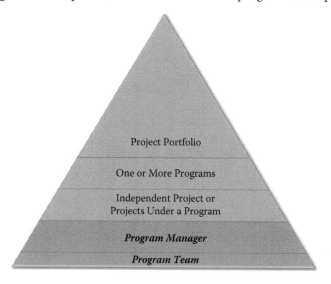

FIGURE 1.3
Recommended program structure.

Those less familiar with a project management environment may confuse programs with project portfolio management.

Portfolio management as defined by PMI (2013a, p. 178) is "the centralized management of one or more portfolios to achieve strategic objectives," and a portfolio is defined as "projects, programs, sub portfolios, and operations managed as a group to achieve strategic objectives" (PMI, 2013a, p. 178). While there are some similarities between program management and project portfolio management, there are also some major differences. The project portfolio and portfolio management are concerned for the overall corporation or organization and should have oversight of all the programs.

The program manager is concerned only for his or her program and the projects that are subordinate to the program. The benefits realization and the strategic benefit are focused on this one program and what the program delivers to the organization or corporation. Alternatively, the portfolio manager would assess all programs and whether or not a program had to be eliminated; the portfolio manager should be able to make an unbiased assessment.

HISTORY OF PROGRAM MANAGEMENT

Ancient engineering and architectural feats may have included program management principles. The Bible details building King Solomon's Temple and his palace (1 Kings 5:1–7:51, King James Version), around 966 BC. The passage includes supplies, dimensions, descriptions of the exterior and interior, where the provisions came from, and a listing of resources. Additionally, the order in which the construction took place is listed. The time frame to build the temple was seven years, while the palace took thirteen years. Solomon declares, "I purpose to build a house unto the name of the Lord my God" (1 Kings 5:5, King James Version). The temple was built to glorify and thank God (1 Kings 5:5, King James Version). From the description there appears to be a strategic purpose, several projects (loosely defined) that are interrelated, and a benefits realization.

Prior to Solomon, the Egyptians built great architectural structures such as the pyramids. The Egyptian pyramids date from 2686 to 2125 BC. The Egyptians appeared to have used some program management concepts, including the management of resources, scope, and quality, and integration management, interrelated projects, and benefits realization for the Pharaoh (kingdom/corporation) (Brier, 2002). These structures were built as a burial site that would assist the deceased to transition to the afterlife, as is the belief of the ancient Egyptians (Brier, 2002).

The Inca Empire, 15th to 16th century AD, exhibited program qualities while building the road systems (Hyslop, 1984). The road system was characterized by being very straight, with major obstacles, and having several different types of buildings at junctures (Hyslop, 1984). Some of the buildings were military, religious, or political in nature. There is also evidence that lodging existed for those who traveled the Inca road system, and lodging was within a day's travel or less (Hyslop, 1984). The speculation is that the road systems met the Inca elite's communication and military needs (Hyslop, 1984). One may extrapolate that the Inca Empire used relatively advanced engineering principles for the road systems that in turn would have required some aspects of program management. The Inca's program management may have included resource, integration, multiple interrelated projects, benefits realization, and milestone management, but did not include cost and scheduled management.

Evidence of program management throughout the ages has been documented by those in power. King Solomon, the Pharaohs, and the leaders of the Inca Empire were wealthy and powerful (Brier, 2002; 1 Kings 5, King James Version; Hyslop, 1984) and did not demonstrate concern for cost and schedule. The Program Standard states that "programs and projects deliver benefits to organizations by generating business value, enhancing current capabilities, facilitating business change, maintaining an asset base, offering new products and services to the market, or developing new capabilities for the organization." The pyramids, temples, and road systems for the Inca were to the glory of God or the person in charge of construction (Brier, 2002; 1 Kings 5, King James Version), but there was no thought to benefit realization.

The beginnings of modern project and program management were recognized in the late 1950s and early 1960s (Simpson, 1970) with the development of the Project Evaluation and Review Technique (PERT). PERT, a critical path method analysis, was developed for the Polaris submarine program. The program was so large that a computer-based system to track the development of the program had to be developed. The U.S. aerospace industry, the U.S. Department of Defense, and large U.S. construction companies drove the program management discipline during this era.

The focus appeared to be on improving profitability while developing new technology (Simpson, 1970; Thomas, 2000).

Of note, in this book the PMI definition for program will be used (PMI, 2013b, p. 166).

PHOTOGRAPH 1.1
Complexity is like a building painted on the side of a building—there is more to complexity theory than what is initially perceived on the surface.

2

Complexity Theory

PHOTOGRAPH 2.0
Non-linear architecture.

APPLICATION OF COMPLEXITY THEORY TO PROGRAM MANAGEMENT

Programs by their inherent nature are complex. The program manager must communicate, just like the project manager. The difference is whether the focus of the communication is strategic or tactical. It is essential to understand that programs go beyond the boundaries of a project, and hence need a program manager who is willing to work within the confines of the program, while at the same time work outside the confines of the program. To achieve this difficult goal, one available and recommended tool is to understand and leverage complexity.

Worldwide there are situations that programs face daily. The first example is with regard to the recent natural disasters that impacted Japan. No one could have predicted that two natural disasters would happen sequentially in such a short period of time. Linear thinkers usually consider disasters as isolated events, and the tragedy in Japan showed that sometimes natural disasters can happen one after another. A devastating earthquake hit, followed by a life-shattering tsunami which quickly began a cascade of events at the Fukushima nuclear reactor plant. This was not a planned program, and time will tell how effective the ensuing program was. Countries throughout the world are faced with natural or manmade disasters on a regular basis. In the last two decades, the United States has faced two that have been studied extensively. These two situations, along with others, will be used throughout the book as striking examples of complexity theory.

A second example is when the U.S. Federal Aviation Administration, after the attacks of September 11, 2001, ordered all air traffic in the United States to be grounded immediately. There were no processes or central command for this effort. To date, this is the best documented effort of complexity theory. Each control tower had to work its own area, coordinating with other control towers, airspace coordinators, and even international aviation administrators to ensure planes landed safely and/or were not being hijacked. Some planes were even diverted to other countries or had to be diverted back to their destinations. This was all done without a central command. There was no program manager to call the shots, yet the ultimate goal of landing all the planes was achieved in a remarkably swift period of time.

The third example is the British Petroleum (BP) oil crisis in the Gulf of Mexico. As the crisis grew, many vessels participated in bringing the disaster under control—too many for one central command to ensure safety for all of them. Complexity theory was in play during this crisis as well. The U.S. Coast Guard command along with BP leadership decided on an entirely different command structure. A group of vessels would be assigned to do its work. This team was independent but did affect the rest of the program. The vessels were very close to each other within the same group and between groups. There were massive tankers and fishing vessels. All had to be cognizant of the environment yet independent of each other.

These three examples are natural or manmade disasters, yet each represents a series of programs that became linked to a predicating disaster or disastrous event. Some would argue that these are not programs because they do not follow the life cycle of a program or they are not projects tied together for strategic reasons to decrease the cost to the organization or company. In Japan's natural disaster, there was not a business case written, but the projects were put together to save the infrastructure of a nation and its people. During the September 11, 2001, grounding of aircraft in the United States, the multiple projects were brought together as a program to minimize the acts of terrorism, minimize human casualties, and minimize the fear of a nation. Finally, the BP oil crisis may have had a business plan once the crisis started. However, there were many business plans written depending on the perspective. There was one done by environmentalists, one for tourism, one by the several states involved, one by the United States, and the list continues. For the purposes of this book, the actual cleanup effort was a program because of all the projects brought together to strategically clean the disaster.

Complexity helps the program manager connect projects together into larger programs that can impact an organization. No one will dispute that the impact of the disasters in Japan, the events of 9/11, and the oil spill in the U.S. Gulf had long-reaching implications that impacted the world. Each of these events parallels the view of complexity that the beating of butterflies' wings in Tokyo could result in a hurricane off the coast of Florida. Nature is unpredictable and complex, and in order to address these kinds of situations a complex view is required that can be deployed when the situation demands. Thus, complexity is about trying to connect things that appear unrelated in a manner that can make sense to human perception.

COMPLEXITY SUB-THEORIES

Understanding complexity theory and its sub-theories helps the program manager navigate through the myriad of issue, risks, and communication dilemmas on complex programs. According to complexity theory, those situations that occur on the fringes normally cannot be dealt with by the program manager, and they shouldn't be; however, the program manager

PHOTOGRAPH 2.1
Ancient meets modern. Note how the castle is together with the satellite dish.

should equip the program team with the skills, tools, and templates to deal with these situations.

In many cases, the program manager needs to navigate through many different challenges. In some cases traditional thinking will be sufficient, but in other cases a more modern, complex approach is necessary for success. Further still, sometimes advanced technology needs to be connected to achieve the best results.

History of Chaos Theory and Complexity

Chaos theory was initially viewed as a hobby among mathematicians and scientists. It was not considered as a real-world application, because human systems were all based upon systems of order. Anything that was out of order was considered to be ineffective and inefficient. Edward Lorenz, a meteorologist, was able to explain the concept of the butterfly effect theory which made some aspects of complexity theory easier to understand. A simplified illustration of the butterfly effect is when a bird flaps its wings in Florida, and this creates a minute disturbance in the atmosphere and creates a hurricane in New York. In the case of the bird in Florida, this minute disturbance may have a drastic change on the weather conditions in North America. This atmospheric disturbance, although small in relative force, may be sufficient to create a hurricane or it may prevent a hurricane where one should have started.

Scientists would have originally called this noise or randomness, and many would have attempted to discount these forces as irrelevant due to their relative small impact when compared to other atmospheric forces. Lorenz found that the atmosphere never reached a state of equilibrium; it is always in a state of chaos. Yet when plotting the atmospheric conditions, it did always plot as butterfly wings or owl eyes. This plot showed that there was some degree of order in the randomness (see Figure 2.1).

What started as a mathematical hobby has turned the world of chaos into one that can sustain order. Lorenz's attractor equation was able to demonstrate order in the world of chaos. In essence, the atmosphere disturbances were drawn to areas or attractors. It appeared as order in what previously was thought to be randomness. As a meteorologist, his work went ignored by mathematicians and others in the scientific community.

In time, scientists would go back and realize that Lorenz was truly working in the right direction with his work in chaos. Eventually, chaos theory was used as an explanation in several areas of sciences for many years before it was even noticed by the world of business. The realm of academia,

FIGURE 2.1
A graphical mockup of Lorenz's Butterfly Attractor model.

and the world of business, is often slow to accept academic theories, and even in academia, it is difficult to make the leap from one area of study to another. Chaos theory was accepted very slowly, and it would take a decade or more before enterprising program managers and academics in the area of project management would see the correlation.

Chaos theory would have to be further distilled before it would be accepted by program managers. The theory would shift into various different areas of study as it matures. This is no different than what is seen with mature professions such as medicine. There are many different specialties including cardiology, neurology, ophthalmology, and so forth. The human body is probably one of the most complex systems, and even skilled surgeons do not know exactly how each person will react to surgery. Great care can be taken prior to surgery, but there are no guarantees that the procedure will be successful and without unforeseen complications. Despite careful cataloging and study of anatomy, each person is different inside.

Evolution of Chaos Theory to Complexity Theory

Chaos theory has moved past the realms of math and science and has moved into social sciences and business. The butterfly effect can be applied and used effectively for distributed programs (Samoilenko, 2008). The butterfly effect is the understanding that all forces are connected. Taking this to program management, when a program is moving forward, it is best to try to put all the forces toward the same direction. Just as the flapping wings of a butterfly in Japan can be a contributing force to the creation of a hurricane in Florida, even a small impact can have a great effect when magnified over time and distance.

A single leader who can motivate each individual in a program and its component projects can assist in creating a controlled hurricane that can achieve complex tasks. Too often people do not realize that even small contributions can build to create something larger than their individual parts. A leader who can motivate and offer small praises that matter can help move a project forward. The more that a program manager, particularly one who operates virtually, can harness this kind of organization, the more effective he or she will become.

Complexity theory acknowledges that humans by nature when living or working together are an open system. Complexity theory acknowledges that there are parts of the system that cannot be explained but recognizes that there is normalcy in the randomness (Byrne, 1998). Western thought seems content to understand the universe as a discrete system rather than a holistic interconnected system. By doing this and examining how a single ant works independently, the dynamics of the colony are not explained. Explaining how the human eye works does not explain the interrelationship of the glands, brain, heart, blood, and so forth. What happens if one part is out of control? How will one part of the body compensate for others?

How Complexity Theory Is Used Today in Program Management

An effective team can be more effective than an individual, and allowing an individual to plow forward can often drive the team further and faster. Complexity is the manifestation of empowering and delegating tasks to allow individuality to support the greater whole, just as a seemingly unrelated collection of stones can be brought together to form a cobblestone road.

PHOTOGRAPH 2.2
Cobblestone path shows how a team of different stones can come together to create a path while also creating their own form of art.

SUCCESSFUL DEPLOYMENT OF COMPLEXITY THEORY

Many organizations are skeptical of the idea that by working counterintuitively, there is value that can be added to an organization. Organizations are concerned about non-linear thinking because the results are not always directly apparent. Complexity theory offers a bridge to move an organization toward non-linear thinking. In the case of Hitachi America, this company was able to improve customer satisfaction and make for a better experience with its products by altering its view of customer service. Instead of listening to what the customer was asking for, leadership asked "what can make the customer service process better so that customer service can make customers happy?" Hitachi America moved from linear thinking to non-linear thinking because the organization wanted to make customers happy.

Hitachi America was continually having a problem supporting products in the field. Customers would call and request a service technician to fix a problem, and Hitachi America could not keep up with the needs of the customers because operations and maintenance just did not have enough service people in the field. So, Hitachi America came up with a non-linear solution. The company stopped sending service technicians when requested by the customers. Instead, customer service representatives with common troubleshooting training would reach out to customers prior to sending a technician. Through research of common problems encountered with new equipment, Hitachi America realized that many of the problems stemmed from a lack of understanding by the customers. By having a customer service person call the customer, Hitachi America was able to help customers get their equipment working sooner than if they had waited for a technician (Blanchard, 2012).

Blanchard (2012) reported that Hitachi America was able to reduce the number of service calls to customer locations by implementing a system where customer service representatives contacted customers with issues before the service tech was sent. It is important to understand that prior to the change Hitachi was only applying linear thinking—a customer would call and ask for a technician to be sent, and Hitachi would schedule a technician to go to the site. Hitachi America was only applying linear thinking because no one ever researched what the problem really was. A linear organization would continue to hire more technicians to meet the demand. By not listening to the customer, and by having a customer service person contact the client, Hitachi was able to reduce service calls to clients' sites by 33%.

Furthermore, Hitachi moved to make its instructions as clear as possible. Solid and clear instructions eliminate or at least reduce the need to receive the help to have equipment to work properly. Someone should have looked to improve the instructions to reduce the problems, rather than hire more service personnel. For those who disagree, consider the change in documentation in instructions for setting up a home computer. In the past, full instruction books were included with a new computer. Now, almost all information is digital, available on the Web, or common software is already pre-loaded to avoid installation issues.

Ultimately, what is important is that individual customer contact points be improved so that the overall experience is *exceptional*. The result should also be that if the department is not receiving continual positive feedback from customers and fans, then the department can do more. What becomes the challenge is to deploy non-linear thinking to an entire program. It may sound difficult, but if an entire company like Hitachi can make the change, then making a similar type of change in a program should be possible.

If one is serious about trying to deploy complexity within a program, then one should consider doing a complexity self-assessment. Find the tool in Figure 2.2 that can offer insights toward what is keeping complexity out of a program.

Name: Date:

COMPLEXITY SELF ASSESSMENT

Instructions	Expect to spend about 15-30 minutes with this self assessment. Try not to over analyze the question and answer with what first comes to mind. There are no right or wrong answers.

STEP 1
Reflect upon your acceptance of Complexity and consider the following: What linear and non-linear factors do you have in your life?

STEP 2
Reflect upon your understanding of complexity and consider the following: How do you use complexity at work? How do you use complexity in a program?

STEP 3
Reflect upon how you resist non-linear thinking and consider the following: Why do you keep resisting complexity?

STEP 4
Reflect upon how you resist non-linear thinking and consider the following: Why do you keep others from accepting complexity?

STEP 5
Promise yourself to change and to take steps to deploy complexity within a program. Make a plan to implement complexity in an existing program.

FIGURE 2.2
Complexity self-assessment.

3

Course Materials #1

PHOTOGRAPH 3.0
Ocean meets the forest.

CLASSROOM MATERIALS

Case Study #1—Leadership, Communication, Complexity

You have been hired as a new Vice President of Retail Programs for a growing mid-sized company that specializes in the manufacturing of commercial and consumer electronics. The company has a number of issues, but the owners have decided to bring in a new person to develop the retail side of the business. The prior vice president could not focus on both the retail and the commercial sides and given his connections with military contracts, it was decided that he be the Vice President of Commercial Programs. The owners recognize that the competition is improving and the internal retail processes are slowing the growth potential of the company. In order to enact sweeping changes throughout the organization, they are restructuring, and you will need to lead this new program as well as work with the other department heads to make sure that you receive your portion of the shared services.

Company Details

The company is called Grace Electronics, and it is a manufacturing company of small commercial and consumer electronics. Grace specializes in high-end, highly reliable, and cutting-edge-type items. The company has a 40-year tradition of producing reliable items and has a good retail distribution through higher-end electronics and home goods stores throughout the United States and Europe. The company does some business through online retailers but finds that the bulk of its business is still through traditional storefronts. The business also has a large commercial side, but that is being handled by a different program manager so the details here will focus upon the retail side of the company.

The company is based in the San Francisco area of the United States. There is also an office in London, UK, for sales, and there is an East Coast office that handles sales, distribution, as well as customer service. The San Francisco facility houses manufacturing, research, and development, U.S. West Coast sales, marketing, supply chain, distribution, and now the reverse logistics department. All manufacturing is done in the United States, but many components are purchased overseas.

The average retail cost for its items is around $45, and the company moved 1,000,000 units last year and expects to do 5% more this year. The company sells through its distribution network items at an average cost of $20 per unit. The total revenue last year was $20 million.

Returns have been steady at 1–2% per year, but there have been recent issues with poor instructions that have been included with products that have increased calls to customer service by 50%. Returns are already trending at 3–4% this year. Customer service consists of two people who are often overwhelmed by calls, and lost or dropped calls have been trending closer to 25% on some days, while last year's average was around 3%.

The last study of returns showed the following internal costs:

Testing costs: $5
Repair costs (when needed): up to $7
Administration and shipping costs: $5

The processing of returns takes 2–4 weeks and has the following breakdown of times: Testing takes an average of 3 days, repairs (when needed) take an average of 6 days, administration takes an average of 5 days, and shipping takes an average of 7 days. The process time for returns is advertised on the website as 2–4 weeks but the UK sales office is advising that returns to Europe are taking longer than the reported times.

The UK sales office moved 50,000 units last year (first year of operation), but it is reporting (unofficially) 2% returns (1,000 units). The UK sales office is verbally advising customers that a return could take up to 6 weeks based upon past experience. Customers making UK returns report that they have had to pay duty for returned goods from the United States.

The sales department is considering opening a new office in Germany within 6 months to address some of the additional demands in Europe. This office would include a customer service office that would handle customer service issues for Europe.

Grace Electronics has recently outsourced its documentation to a company in India. Outsourcing the documentation and instructions reduced overhead by one person and yielded a savings of $45,000 per year. Not everyone has been happy with this change, but the company owner feels that the savings was worth it. This is something that will need to be looked at as documentation and instructions are important to the retail items being distributed.

New inventory controls have been reported to be confusing as the documentation for the inventory controls is unclear. The person who did the documentation used to help on these kinds of policy changes, but since that job elimination there has been no one to pick up that extra work.

The Logistics Manager, Jason Summers, is new (started a few days before you) because the last person in this position left over a dispute with the owner over the new inventory controls. The new logistics manager is not familiar with the new procedures and is still trying to become comfortable with his new role, as his prior position was handling mostly international sea freight. Note that the last study of inventory control showed that inventory accuracy has decreased from 98.50% to 92.75%.

Customer service is located in Florida due to the majority of the sales being on the U.S. East Coast. It was under the sales group prior to your arrival and prior to the creation of the reverse logistics group. All returns are routed to the manufacturing facility in San Francisco; however, returns are accepted at any location, but this increases costs and handling because everything needs to be re-routed. The additional handling has increased costs by an estimated 25% and has pushed up the average time for the administration of the returns.

The plant manager, Rick Townson, wants to improve returns due to pressure from marketing/sales due to customer complaints. The plant manager has pressured the supply chain manager to improve the transportation accuracy, because he feels that many of the complaints are due to the items not arriving on time. The third-party logistics company that handles the freight to consumers and to the distribution network is based in Seattle, Washington, and had a delivery accuracy of 98% last year.

The plant manager has mentioned to you that he feels that things could be improved by offering an incentive (say a few extra dollars per repair) to the production employees who repair returned goods. He feels that this would be a better alternative than outsourcing. Rick resisted the change in the documentation process, and the owner finally had to step in to implement the change to make sure that the cost savings were realized.

Outsource Options

Prior to your arrival, there was an initial bid solicited regarding outsourcing in the returns process. Some people feel that an outsourced company would be better than trying to force the changes within the organization.

Company #1

This reverse logistics company is based in Iowa, and it offers a reduced processing time of 1–2 weeks. Grace feels that this improvement will assist in improving the company's position in its market niche. The company feels comfortable that it will be able to achieve this, and Grace would be willing to market this along with the company to assure clients of this improvement. The company has bid at $15 per return (testing and repair) but would need initial funding to pay for training and startup of this operation. The reverse logistics company would be able to train key persons in

Grace Case Study Comparison		
Questions	**Answer/Reasons**	**Costs**
Reasons to outsource		
Reasons against outsourcing		
What else do you feel needs to change?		

FIGURE 3.1
Grace Electronics comparison matrix.

the repair process and would then be able to purchase supplies to support this operation.

Company #2

This reverse logistics company is based overseas in Taiwan. It feels that it would be able to handle the repairs for $11 per unit (testing and repair). The company feels that shipping costs might increase but will be charged as a pass-through cost. The reverse logistics company would need some initial training support, but this could be done by sending over a person to teach its operation for the repair techniques for the various items. The reverse logistics company feels confident that it can handle the returns in 3–4 weeks and would be willing to review this time frame after 6 months to see if they can improve it.

Both bids are still valid, and both companies are contacting you for an update to know if they will have the business moving forward (Figure 3.1).

Case Study Questions

1. Given the organization, would you consider outsourcing the returns process?
2. If you choose to outsource, which company would you select? Explain why you would use a particular company.

3. If you choose not to outsource, explain how you would better organize the groups currently handling returns.

4. What other changes would you make to the organization? Why? Support your reasoning.

SECTION QUIZ

This section is separated into four parts. Section 1 has multiple choice questions, Section 2 has true/false questions, and Section 3 provides answers to Sections 1 and 2.

Section 1

Multiple Choice

1. What complexity sub-theory is behind the popularity of the Internet's social media?
   a. Butterfly effect
   b. Six degrees of separation
   c. Adaptability leadership
   d. Transformational leadership

2. Program management is defined as:
   a. Strategic
   b. More than one project to deliver value
   c. Complex by nature
   d. All of the above

3. Complexity theory has the following characteristic:
   a. Non-linear
   b. Simple
   c. Linear
   d. Finite

4. Complexity theory acknowledges the following: (select more than one)
   a. Non-linear thinking
   b. Humans are open systems

c. Acknowledges humans react unpredictably

d. Programs are linear

Section 2

True/False

1. A butterfly that flaps its wings in Japan may cause a hurricane in Florida is commonly referred to as the butterfly effect.
   a. True
   b. False

2. Program managers should ignore the fringes of chaos; it really does not matter.
   a. True
   b. False

3. Effective program managers combine linear and non-linear thinking and apply the type of thinking in appropriate situations.
   a. True
   b. False

Section 3

Answer Key

Section 1

1. A
2. D
3. A
4. A, B, C

Section 2

1. True
2. False (Program managers should be aware that chaos happens on the fringes but that they cannot manage all the chaos. Program managers should never ignore the chaos but empower the team to handle the situations.)

3. True (Not everything on a program is non-linear. There are traditional program management issues/risks/costs/quality that are done in a traditional linear fashion.)

DISCUSSION QUESTIONS

Short Discussion Questions

What is chaos theory?

What is complexity theory?

What is the difference between chaos theory and complexity theory?

Explain the history of chaos theory.

Explain the history of complexity theory.

Compare the histories of chaos theory and complexity theory.

Define program management.

Give an example and explain a successful program.

Give an example and explain a less-than-successful program.

Explain program management and how it is different than project management.

Explain how complexity theory can assist with program management.

Give some examples of complexity theory being applied within a program.

Explain some of the complexity sub-theories.

Explain how complexity theory can be deployed within a program.

Long Discussion Questions

Define and discuss the evolution of chaos theory into complexity theory. Give examples of each and indicate the significant impacts of the history of both.

Give an example of a successful program that was able to leverage complexity theory in some manner. Explain the importance of human factors in the deployment of complexity.

Define, explain, and critically evaluate the importance of complexity sub-theories as compared to chaos and complexity theories. Support your position.

Research complexity theory and offer recommendations on how complexity theory could have been deployed successfully in existing or previous programs. Defend your answer and use support to show how complexity theory would have been more successful than another methodology.

Define program management, offer a potentially successful linear method to manage the program, and then offer a potentially successful non-linear method to manage a program. Compare and contrast the two positions, and offer your recommendations on both.

SUMMARY

Seasoned program managers realize that a complex program has so many parts that it is impossible for even the most detailed micro-manager to control (Jaafari, 2003). As projects have become larger and more diverse, programs have evolved to being even more diverse and complicated. A single manager does not have the time in a day to manage every aspect and element of a program. This has forced individuals to realize that there are limits to control; however, they must then deploy new methods to organize the program. The program manager must become more creative in monitoring the fringes of complexity or chaos.

Some program managers who have coped with these situations have often leveraged conventional linear thinking with practices such as delegation, empowerment, or stewardship. The difference is that these named management theories are about taking action or controlling singular actions or tasks. Employees can be delegated the authority to handle tasks; employees can be empowered to make certain decisions, and a leader might position himself or herself as a steward of the organization; however, all of this falls short of complexity. These theories and management ideas are important and effective; however, the difference between complexity and these others is the role of the manager. In all of these prior theories, the leader was always the final authority in the paradigm. The program manager was the only authority, and the program manager set the rules and his or her rules were the law (Bass, 1990).

A linear program manager is one who followed the example of the leader of an artillery team. In a linear organization, artillery fire was handled by a team of individuals who were focused on loading and firing the artillery. The artillery team was directed by a single individual who would monitor and gauge the effectiveness and accuracy of the projectile. The forward observer would then direct the team in order to modify the direction of the fire. The forward observer's decision was then executed by the artillery team (U.S. Army & Marine Corps, 2007). This is the template for a classic linear program. In this case, the program manager assumes the role of the forward observer, and the rest of the project team is responsible to load, fire, and aim the artillery.

Military equipment has advanced to the point where targets are identified and smart missiles fire upon a target. There is no longer leader interaction to direct this kind of fire. This fire and forget technology is akin to complexity theory. Targets are identified by technology regarding which are friend or foe, and then the individual, assisted by technology, fires upon the largest threats or targets where they are most likely to eliminate. Program managers must learn to turn loose their teams upon problems/tasks/duties and then allow them to solve them and move on (U.S. Army & Marine Corps, 2007). The overriding assumption is that individual projects within the program are completed and not left incomplete. Leaders need not follow up with others on the team to know if assigned tasks are done. Assigned tasks are completed on time, and the team moves on to the next task. This is the new paradigm that embraces complexity.

What disturbs most people is that this kind of thinking makes a team appear to be in total chaos. Since individuals are expected to do their best work for the program, some work teams may experience a catastrophic failure of some sort; however, they are allowed to resolve it on their own. Making the team handle its own failures will force the team to accomplish this task quickly and efficiently (Hass, 2009). This appears to go contrary to standards suggested for linear program managers who want to be consulted with problems in order to offer their expert input. The weakness with that is that problems cannot be addressed as quickly as they arise, because people feel the need to explain the options. This gives the appearance that a program is totally out of control, because teams are not alerting the program manager of potential issues. Furthermore, when there is no formal reporting structure to ensure that all matters are handled, people feel that they are being excluded; however, one must foster an environment where professionals are addressing problems and moving on to the next task without intervention (Hass, 2009).

To sum up, to be successful with complexity, the program manager must completely understand the theory and implementation within the scope of program management. The program manager must also make sure that he or she has the right people in the right spots, because having a linear thinker who is always waiting for direction will be problematic on many levels. The program manager must understand the integration of the processes and must have self-confidence in his or her ability as a leader as well as confidence in the people to *always* do the right thing (Hass, 2009). Once the right people are in place, then the program manager is ready to embrace the ability to allow *chaos* on the project. If the program manager is not willing to let go to this level, then it is not recommended to move toward complexity.

PHOTOGRAPH 3.1
A sunset offers a complex view of the different wavelengths of light.

4

Leadership in Program Management

PHOTOGRAPH 4.0
Archway of knowledge of leadership.

INTRODUCTION TO LEADERSHIP

Leadership has been an evolving art since people have recognized the importance of organizing, be it for mutual defense to preserve a people or to organize people into an efficient work group. To understand leadership, one must review the historical context of leadership and the way that leadership has evolved to the modern day. The history of leadership can be viewed with a perspective of eras and theories. In order to capture the different leadership trends throughout the ages, different leadership eras have been defined.

The eras are: Early Leadership, Industrial Leadership, Modern Leadership, and Post-Modern Leadership. This review of the evolution of leadership is based upon a combination of leadership eras proposed by Chemers (1995) and those described by Bass (1990). Chemers (1995) proposed three periods of leadership: the Trait theory period from 1910 to World War II, the behavior period from World War II to the 1960s, and the Contingency theory period from the late 1960s to the present (Chemers, 1995, p. 83). The Early Leadership and Post-Modern Leadership groupings are defined by leadership research from Bass (1990), while Industrial and Modern Leadership are based upon Chemers' model. Table 4.1 defines these different eras of leadership.

Early Leadership

Authoritarian leaders dominated the Early Leadership era. Authoritarian leaders ruled through right and privilege, based upon the leader's duty and obligation (Bass, 1990, p. 3). Early leaders emerged for mutual protection. Egyptians used leadership to construct pyramids, and Romans used it to control their vast empire (Dessler, 2001, p. 29). Religious leaders, chiefs, and kings were not only the heads of early government, but they also

TABLE 4.1

Eras of Leadership

Leadership Era	Time Period	Leadership Theory
Early Leadership	Early civilization to 1920	Great Man
Industrial Leadership	1920–1948	Trait
Modern Leadership	1949–1984	Contingency
Post-Modern Leadership	1984–Present	Transformational

served as role models who embodied the culture of early society (Bass, 1990, p. 3). From the early days of civilization until the industrial age, the existing theories of leadership revolved around the Great Man theory.

The Great Man theory explained leadership as a right of birth, station, or caste. In this theory, leaders are born, not made. According to the prevailing leadership theory of this era, individuals were born with leadership ability, and those of the highest stations of society were the best leaders. Bass and Galton both studied the hereditary backgrounds of men in an attempt to defend this theory (1990, p. 38). Woods studied 14 nations and their rulers over a period of ten centuries and found that the kings and their kinship were powerful and influential, and he linked this to their natural abilities at birth (Bass, 1990, p. 38). The industrialists of the early twentieth century did not all come from the highest class of society. In order to explain the emergence of these industrial leaders, a new theory evolved based upon the belief that certain individuals had inherent leadership abilities. Much like a musical protégé who displays great musical aptitude at a young age, leadership was thought to be a similar quality.

Industrial Leadership

Modern Leadership theory began to emerge in the early Industrial Age, when many individuals deserted their roles in agrarian society and were forced to seek income in the urban centers of society (Jacques, 1996). This created a displaced workforce of individuals longing to return to their rural roots. Workers migrated to urban centers in order to make enough money to buy a farm and return to an agrarian lifestyle.

During the Industrial Leadership period, workers and managers clashed as the expectations of labor grew.

> A government study shows that state troops were called out to calm unrest nearly 500 times between 1875 and 1910. This figure does not include the actions of private mercenary groups, such as the Pinkertons. The reigning mood during this period was one of terror. (Jacques, 1996, p. 60)

As the industrial era progressed, people began to migrate voluntarily from these farms to seek their fortunes elsewhere. These individuals expected to work daily in factories in order to receive payment for their physical production. During this transition, a decided shift occurred from inclusion in the community to exclusion from the community. When

America shifted away from a society where one had to be successful within the community, the country moved toward an industrial, opportunistic society (Jacques, 1996, p. 26).

During this time, there is the emergence of Trait theory, which becomes the dominant leadership theory of this era. Trait theory is an extension of the Great Man theory, because in both theories, leaders are born. Trait theory states that traits occur naturally regardless of station or caste. According to Trait theory, leaders are born with certain leadership talents regardless of social status. Trait theory ascertains that leaders' characteristics are different from those of non-leaders (Kirkpatrick & Locke, 1995, p. 134).

A study in 1948 by Stodgill reviewed over 120 traits in an attempt to discern a pattern to Trait theory; however, this research was inconclusive (Chemers, 1995, p. 84). Without conclusive evidence to support the theory of leadership as trait-driven, it brought a shadow of doubt toward this theory of leadership. Stodgill's inconclusive evidence effectively ended the Trait theory era. As people began to search for other leadership explanations, the concept of Contingency theory emerged.

Modern Leadership

By 1949, workers had accepted their wage earner roles as agrarian workers were on the decline. Individuals who expected to live and die as wage earners mark the Modern Leadership era. Individuals began to expect that they would live their lives working for a company as a wage earner (Jacques, 1996, p. 96). Because of this change individuals expected fair treatment, and leaders could no longer rule with an iron fist, fearing that followers would strike, or worse. During this time of the cold war between capitalist and communist societies, some of the predictions of Marx and Engels began to emerge:

> Thereupon the workers begin to form combinations (Trade Unions) against the bourgeois; they club together in order to keep up the rate of wages; they found permanent associations in order to make provision beforehand for these occasional revolts. Here and there the contest breaks out into riots. (Marx & Engels, 1978, p. 480)

In response to these actions by labor, the government began to guarantee the rights of individuals who toiled in organizations.

In the Modern Leadership era, management and labor had to learn to communicate and work together effectively in order to maintain order and peak productivity. In 1954, Gibb proposed that leadership is actually an interactive phenomenon emerging from group formation (Bass, 1990, p. 40). This new concept of group theory leads to the development of the contingency theory of leadership. Contingency offers leadership as a function of task-oriented and relationship-oriented situations, which relate to group performance (Bass, 1990, p. 40). Fiedler's Contingency theory dominates much of this era. Although Fiedler's Contingency model was subject to considerable scrutiny, his personality measure of the esteem for the least preferred coworker offers a relationship-motivated perspective on leadership (Chemers, 1995, pp. 86–87).

Post-Modern Leadership

The Post-Modern era of leadership offers the model of effective leadership. Leaders of this era were valued for their effectiveness and knowledge, rather than their abilities to relate to groups. During the transition from the Modern Leadership era to the Post-Modern Leadership era, the overall education of the general population increased. This rise in educational service supports the idea that the knowledge worker was on the rise.

In 1920, 9% of the population engaged in knowledge and education services, while in 1995 the percentage had risen to 29%. Furthermore, in 1975, the percentage had increased to 50% (Bass, 1990, p. 881). From 1982 to 1992 the number of management schools had grown from 545 to 679, representing a growth of 23%. The number of MBAs granted from 1982 to 1992 had gone from 60,000 to 80,000, a growth of 33% (Nohria & Berkley, 1998, p. 201). During this period, new data began to emerge regarding effective management. Organizations of the information age that support collectivism, tradition, and androgyny are proving to be more effective than organizations that favor masculinity and autocratic behavior (Bass, 1990, p. 912).

This era of leadership saw the emergence of transformational leadership. Transformational leaders transform followers into leaders. Bennis (1994) was able to identify 90 transformational leaders and found evidence to support the following elements of leadership: competence to manage attention and meaning, the communication of a possible vision, and the empowerment of those working toward a collective goal (Bass, 1990, p. 53). Bass (1999), Cohen and Tichy state that transformational leadership is a "behavioral process capable of being learned and managed. It's a leadership process that is systematic, consisting of purposeful and organized search for changes, systematic analysis, and the capacity to move resources from areas of lesser to greater productivity" (1986, pp. 53–54).

TRANSFORMATIONAL LEADERSHIP

"The force creating success is the leadership that creates a fusion of process and entrepreneurship" (Andraski, 1998, pp. 9–11). This quotation captures the essence of the style of the modern transformational leader. The entrepreneurship of the leader is the ability to have a vision of a new organization that can evolve beyond the current organization. According to William Kieschnick, former CEO of the Atlantic Richfield Corporation, the biggest problem faced by a multibillion-dollar corporation is the ability to infuse an entrepreneurial spirit at every level of leadership (Bennis & Nanus, 1997, pp. 208–209). This vision consists of a "central concept 'system' which embodies the idea of a set of elements connected together which form a whole, thus showing properties which are properties of the whole, rather than properties of the component parts" (Checkland, 1999, p. 3). These values are the foundation of the organization and must become the shadow of the leader. The communication of the vision becomes the voice of the leader. In summary, vision, value, and voice define a leader's style.

Vision

A leader is an individual who has a vision of a new reality and the internal motivation to achieve the goals of this desire. This passion of vision offers inspiration, empathy, and trustworthiness (Bennis, 1994, p. 140). When a leader has a compelling vision of a new world order, that leader can display "courage under fire" (Useem & Harder, 2000, pp. 25–36) in order to reach that distant goal. This new view must not only be understood by the leader, but must also be communicated to others who are touched by the vision. Vision within a company involves all leaders, no matter at what level (Collins & Porras, 1997, pp. 1–2). The leader must shape the organization in a creative manner in order to accomplish a new order of reality. A good vision is one that has the capacity to create and communicate a new paradigm, context, and frame that invoke commitment and clarity of action (Bennis, 1994, p. 378).

Values

Core values are important to a leader. "To be able to manage oneself, one has to know: What are my values?" (Drucker, 1996, p. 175). What beliefs the leader finds important are critical to becoming effective in an organization. Leaders must translate values into action (Badaracco, 1998, p. 91). Leaders must have values that reflect the organization while reflecting their own spirit. Useem and Harder (2000) found that the most important values of leaders were strategic thinking, deal making, partnership governing, and managing change. According to John Sculley, as cited by Bennis, "leadership revolves around vision, ideas, direction and has more to do with inspiring people as to direction and goals than with day-to-day implementation" (1994, p. 139). Once the group shares these values, then the leader must maintain the values of the organization.

The one value that is difficult to maintain is the value of change. Most organizations do not value change, and some organizations fear change. An example of this resistance can be seen by the reaction that employees have toward outsourcing. "Outsourcing is frequently accompanied by employee resistance. For most hourly employees and many managers, outsourcing is synonymous with job loss or change" (Useem & Harder, 2000, pp. 25–36). Rather than view change as an entrepreneurial opportunity, most organizations view change as an "evolution and revolution" where a "management sprawl leads to a series of confrontations among the layers of management" (Brown, 2000, pp. 52–55). Leaders must attack change head on, through aggressive educational training and solid assurances to their followers. Transforming leaders must embrace change so that it becomes another accepted process, akin to promotions and the development of new managers. The transforming leader commits people to specific actions. A transforming leader creates leaders out of followers and can convert leaders into active agents of change (Bennis & Nanus, 1997, p. 3).

Voice

Voice is the ability to persuade others to the new calling of the organization. Leaders who communicate a vision and values offer the followers a road map that will give guidance regarding the decision-making process.

Since leaders will not always be present, followers must be prepared to make critical decisions based upon the vision and values of the leader. "The best leaders are perceived as having highly developed interpersonal skills. They are warm, open, and forthright, and their attitude toward employees is characterized by words like 'empowering,' 'supportive,' and 'benevolently paternalistic'" (Hemsath, 1998, pp. 50–51). Interpersonal skills become the 'voice' of the leader. These interpersonal skills allow the leader to issue edicts and orders, but they are also the vehicle to offer advice and to support the decisions of others.

Leaders must learn to make a difference by instilling ownership and camaraderie. "Leadership that emphasizes negotiation and coordination over authoritarian strategies to facilitate the internal and external communication" (Kent-Drury, 2000, pp. 90–98) will create the successful organizations of the future. Leaders must successfully listen to the voice of the organization, and to then echo these ideas in their own organizational voice. Giving voice to others is important to any organization that is willing to learn and change (Heifetz & Laurie, 1998, pp. 190–191).

As discussed, voice is important to all communication, and vision and values can be summarized into the meaning of communication. If communication is meaningful, then it has vision and values. Consider the matrix presented in Figure 4.1 for any message that is being sent to a program or a project.

Transformational Leadership and Teams

True leaders understand the value of blending the culture of the organization with their own distinct philosophy. Culture is a construct that contains all the norms of the tribe. Business philosophy is based upon the leader's vision to go beyond the clever spreadsheet and market forecasts to develop new beliefs that motivate employees, customers, and stakeholders. "Leaders are the most results-oriented individuals in the world, and results get attention" (Bennis & Nanus, 1997, p. 26). A leader must learn to value the individual while instilling a successful philosophy upon the business venture. "As long as we regard people in terms of earning power or specific expertise, we do not see their character. Our lens has been ground to one average prescription that is the best suited for spotting freaks" (Hillman, 1996, p. 255). Leaders must learn the "skill of adapting—being able to adjust or fit your behavior and your other resources to meet the contingencies of the situation" (Weiss, 1999, pp. 6–9).

		MEANING				
		HIGH 1	HIGH 2	MEDIUM 3	LOW 4	LOW 5
V	HIGH 1	Total Communication	Good Communication	Neutral Communication	Marginal Communication	Marginal Communication
O	HIGH 2	Good Communication	Good Communication	Neutral Communication	Marginal Communication	Marginal Communication
I	MEDIUM 3	Neutral Communication	Neutral Communication	Neutral Communication	Neutral Communication	Neutral Communication
C	LOW 4	Ignored Communication	Ignored Communication	Neutral Communication	Weak Communication	Weak Communication
E	LOW 5	Ignored Communication	Ignored Communication	Neutral Communication	Weak Communication	Poor Communication

FIGURE 4.1
Meaning and Voice Matrix.

New forms of teams and the distribution of knowledge are being heralded as the innovative culture of the future. Leaders must remain informed about business culture norms in order to perform their duties. The age of the Internet has brought about new techniques and technologies, including asynchronous collaboration and communication, which has become a necessary element of today's geographically dispersed business (Elkins, 2000). Understanding the changing cultural forces that influence customers, employees, and stakeholders is critically important in business growth. The companies that understand the evolving expectations of employees, customers, and stakeholders will flourish. Those companies that can meet these requirements will become successful in the future, while those that fail to do so will shrivel (Duarte & Snyder, 1999, pp. 3–4).

Effective leaders have a business philosophy that is clearly understood and communicated. The philosophy must be composed of articulated values that support the goals of the business organization. "It is assumed here that if core values are neglected in the policy deployment process, it will never be possible to achieve business excellence" (Dahlgaard, Dahlgaard, & Edgeman, 1998, pp. S51–S55). Leaders should strive to stand out in a crowd by understanding the business culture of the organization. Leaders should instill a philosophy that supports their business goals and objectives. Hillman offers the acorn theory as an example of this philosophy of self-expression. The acorn theory states, "that every person bears a uniqueness that asks to be lived and that is already present before it can be lived" (Hillman, 1996, p. 6). Leaders must strive to discover their inner vision or inner voice, and then have the courage to express these thoughts within their business organization. Hillman contests that this inner vision is the discovery of our daimon, which we have been given before we are born (Hillman, 1996, p. 8). Bennis claims that this inner voice is what gives internal direction to great leaders (Bennis & Goldsmith, 1997, p. 24). Direction comes from those who stand apart from the crowd, which means to embody values that may not fit today's fashion. Leaders who stand apart from others offer a different view of the future, a view that others might find appealing.

Transformational leaders must master "the skill of communicating clearly—being able to communicate in a way people can easily understand and accept" (Weiss, 1999, pp. 6–9). Unlike business culture, which is a learned skill, a company business philosophy must be modeled to those within the organization. Leaders who persuade individuals to attain greater productivity will always lead corporations that offer stakeholders greater value.

Leadership and teams revolve around the harnessing of the synergies of organizational commitment, communication, and technology. Leaders who master different forms of communication will have an edge in the virtual environment. Leaders who learn to generate greater commitment to roles and companies will also flourish. Leaders who understand the importance of funneling trust to meet the needs of the program will be more successful (Figure 4.2). When a leader can combine time, belief, credibility, respect, and understanding, then the leader can really build trust in a manner that can influence the entire program.

Finally, the evolving technology must become part of the leadership of the workplace in order to achieve a successful virtual strategy.

In summary, leadership is about influencing others through credibility. In the end, leadership is a highly personal matter, and everyone must decide on his or her best method. Leadership requires a learning strategy and must engage people in confronting daily challenges, adjusting their values, changing perspectives, and learning new habits (Heifetz & Laurie, 1998, p. 197). Regardless of what defines a leader, a great leader will be instantly recognized by others when he or she is in action. Leaders get things done, and programs are all about getting the possible done.

Trust Funnel

FIGURE 4.2
Trust funnel.

PHOTOGRAPH 4.1
Departing archway.

5

Leadership and Program Management

PHOTOGRAPH 5.0
Small castle on a hill.

MEGA-PROGRAM LEADERSHIP

Regardless of the program size or scope, success is always defined as fulfillment of program benefits in a certain period of time while remaining under a certain budget. The theory of constraints explains that scope, schedule, and budget are three important aspects of any program or project, and if one of these elements changes then it will likely have an impact upon another aspect (Guide to the Project Management Book of Knowledge, PMI, 2013a). Many program managers believe that the impact is proportional to the change; however, understanding best practices may offer insight into reducing these impacts. In some cases, the fundamental ideas need to change in a program in order to achieve these best practices, while in other cases these best practices can be superimposed upon a program in order to achieve greater success. The program manager is left to address program culture as each program culture is different. However, a resourceful program manager should try to apply different best practices in order to achieve greater mega-program success. The objective of the program manager should be to apply different best practices to mitigate issues pertaining to scope and schedule through the application of best practices of the virtual program management office (VPMO) and complexity theory (CT).

Virtual Program Management Office (VPMO) Best Practices: Scope Management

For those unfamiliar with the definition of a VPMO, it is a management team that is not co-located and that is responsible for a program or project team (Gordon & Curlee, 2011, pp. 3–4). In a mega-program, the VPMO is tasked with having to manage a large and complex program with multiple component projects and likely with multiple strategic stakeholders. The complex nature of a mega-program increases the responsibility of the program manager who will find that he or she must address all of the needs of stakeholders while having less time for internal people issues.

Program managers of mega-programs need to be masters of time management, and they must learn to effectively manage and maintain trust and positive communication in a VPMO. Since a virtual program manager does not have first-person daily contact with team members or stakeholders, it is important that other methods be used to maintain positive communication and to continually build trust. Research has shown that the virtual environment is the hardest environment in which to build trust (Duarte & Snyder, 2006; Gordon & Curlee, 2011; Tavcar, Zavbi, Verlinden, & Duhovnik, 2005), and a mega-program increases this challenge considerably. Furthermore, communication by phone and other online communication are not as robust as face-to-face communication (Duarte & Snyder, 2006) which often leads to a personal disconnect to the program. Additionally, the increased use of email does not always result in positive communications and trust (Curlee & Gordon, 2010). Without trust-reinforcing contact between the VPMO and program members and stakeholders, and robust communication, individuals can lose touch and people will often feel isolated from the mega-program. Thus, two VPMO best practices are building and maintaining trust and positive effective organizational communication.

VPMO Building and Maintaining Trust Best Practice

Although it is not easy, it is important that a VPMO invest in creating and maintaining trust in a mega-program. Research has consistently supported that trust is an integral part of a successful virtual team (Anderson et al., 1998; Curlee & Gordon, 2010; Duarte and Snyder, 2006; Lipnack & Stamps, 1999; Tavcar, Zavbi, Verlinden, & Duhovnik, 2005). As trust is so important in a virtual organization, one important best practice that has emerged with regard to mega-programs is the establishment of a *federation* in order to unify all those involved with the mega-program. The *federation* concept is creating a participatory relationship for the program team and stakeholders.

This concept makes everyone involved with the program an accountable element in the final deliverables. If everyone is directly invested in the program, then people are more apt to do their best for the success of the program. This would mean making people socially, fiscally, and culturally invested in the program. This participation should include all mega-program stakeholders, because if everyone is seen to have valuable input toward major decisions, people will become more invested. This does not mean that every employee has a voice, because that would become unwieldy in a mega-program; however, one should consider the U.S. federal government method where representatives are identified as the individuals to represent the needs of any particular constituency. This type of representative decision-making process can help manage scope in a manner that allows everyone into the process. It will not guarantee that the agreed upon scope will make everyone happy, but it can help everyone understand what will be done and why.

It is common in mega-programs to create a *federation* brand. A brand might include slogans such as "5989 on time!" or mega-program logo-wear such as shirts, hats, or pins to better define and improve the cohesiveness of the federation or the simple creation of a program goal card. Creative slogans are low-cost options because anyone can repeat a slogan and include it as a tagline for emails; the logo-wear option might represent large investments that might not allow for logo-wear to be possible for all programs. If the budget allows and if the program is appropriately funded, then a logo-wear option helps create the feeling of a *federation*. However, a low-cost alternative to logo-wear would be to create a program goal card that can be distributed to everyone involved. The simple creation of a business card that includes an image of the final mega-program deliverable with the program name and the top goals and/or values of the program that is distributed to everyone involved can create the feeling of inclusion (Gordon & Curlee, 2011). What is important with this route is that the more personalized the card, the more likely it is to create the feeling of inclusion.

Boudreau, Loch, Robey, and Straud (1998, ¶ 13) found that virtual organizations that leverage the notion of a federation are more successful than those that do not. The federation concept as defined by Boudreau et al. (1998) is virtual partnerships, joint ventures, consortia, and other alliances that are managed by a group that are designed to change with a program. A federation may include alliances with other outside organizations or stakeholders involved with the success of a program (Boudreau et al., 1998). The federation concept has been applied successfully to the B-1 Bomber program, which had over 2,000 corporations working together, most of these whose primary interaction was virtual. Other successful corporations that utilize a federation concept include Sun Microsystems, Nike, and Reebok (Boudreau et al., 1998). A federation can help build a community that is focused upon the success of a mega-program. A federation that works together will be able to work out scope details better than an organization that lacks a process that involves all participants with the scope of the mega-program.

VPMO Positive Effective Organizational Communication Best Practice

Since trust and communication are a major part of a successful virtual team (Dani, Burns, Backhouse, & Kochhar, 2006; Handy, 1995), the leaders of a VPMO must establish, maintain, and evolve consistent values and boundaries in the organization and continually communicate those values to all stakeholders. One manner to communicate these values and boundaries is to advertise the successes and failures of the program. Honest success leads to more trust and communication as stakeholders have something positive to celebrate. A successful program will grow new fans when the milestone success of the program is seen by others. The VPMO must recognize that if program success is the panacea for trust, then program setbacks are its nemesis.

Program setbacks, if not properly managed, can undermine trust and communication faster than any other factor. When a program is faltering, people will try to distance themselves from the program (or organization) in order to avoid being associated with the failure. This becomes a challenge for the VPMO, and the best way to meet this challenge is to communicate the failure and explain the implemented solution. If the program is behind, the VPMO must communicate to everyone the strategy to get back on track. The VPMO must explain what scope will be changed or modified in order to meet the schedule and budget and realize the intended benefits. The VPMO must also explain how it will accelerate or decelerate components in order to meet the new challenges of the mega-program. This must be done through multiple communication means because the attitude of "failure is not an option" must permeate the mega-program in order to find solutions rather than be stopped by obstacles (Gordon & Curlee, 2011).

VPMO Summary

Virtual program managers involved in mega-programs are often too busy to focus on the fundamentals; however, Abraham Lincoln said it best when he stated, "A house divided against itself cannot stand." The fundamentals make any program successful, and this is even more important for a mega-program. When program managers forget the basic building blocks of success, then success becomes more elusive. Social skills are just as important as strategic goal alignment and benefit realization. If people do not feel included and do not feel that they are being led correctly, there will be strife, confusion, or worse, conscious or unconscious sabotage. Achieving and maintaining trust and effective communication is the only way to achieve success. When people are involved and invested in a program, then they are willing to work together to find solutions rather than to accept an impasse that slows or halts the program until it is resolved.

Having a rapid resolution process because people want the program to succeed will make people more flexible with regard to scope.

To sum up, if there is time to make status calls and write update reports, there is time to maintain a positive and trusting relationship with stakeholders of the program. People agree that trust is earned and the VPMO must work toward earning the trust of others (MacPhail, 2007; Tavcar, Zavbi, Verlinden, & Duhovnik, 2005) so that when scope challenges arise people strive to find solutions. Communication and building trust must be part of the daily agenda rather than be addressed when there is a problem, because when addressed daily, it will never become a problem. Consider this new paradigm for a successful program: instead of using the yardsticks of scope, schedule, and budget, use the yardstick of *did you feel glad or sad when the program came to an end?* Gladness implies that one could not wait to get out of that program team, while sadness implies that one trusted the group and enjoyed the program (Gordon & Curlee, 2011). If people are sad when the program is coming to an end, then people have worked hard together toward success and acceptable compromises with scope have been made along the way while still preserving the essence of the program. Thus, trust and communication are imperative to the overall successful perception of the final deliverables of a program.

Complexity Theory Best Practices: Schedule Management

Complexity theory has its roots from Edward Lorenz, a meteorologist, who explained that to understand highly complex systems one must take into account all forces involved in the system, and no small element can be ignored because even the beating of the wings of a hummingbird could have an impact on the weather on the other side of the globe. Even modern techniques of weather prediction can only approximate the movement of significant weather. Since this could not be calculated, this was originally believed to be randomness, and many have attempted to discount these smaller forces as irrelevant due to their relatively small impact, but Lorenz found that the atmosphere never reached a state of equilibrium; it is always in a state of chaos.

Chaos theory has moved past the realms of math and science into social sciences and business. The butterfly effect can be applied and used effectively for large-scale virtual programs (Samoilenko, 2008). From a program management perspective, when a program is moving forward, it is best to put all the forces together to work in the same direction. Just as the flapping wings of a butterfly in Japan can be a contributing force to the creation of a hurricane in Florida, even a small impact in a program can have a great effect when magnified over time and distance.

From these ideas about chaos, complexity theory has emerged as the management belief that total order does not allow for enough flexibility to address every possible human situation. People are inherently skeptical of less order because it is believed that it leads to less control. A recent example of where complexity theory worked at a level never before tested was with the investigation of air traffic controllers after the 9/11 tragedy. Once it was known that an unknown terrorist group was hijacking planes to attack buildings in the United States, it became a national priority to have every plane in the airspace of the United States land at the closest airport. Since this kind of crisis had never existed, there was no procedure or process in place to allow this to happen. Researchers wanted to determine the best process or procedures to address this type of widespread domestic crisis were it to happen again. The researchers examined how each set of air traffic controllers managed the situation. In the end, the study concluded that the best way to handle such a crisis would be to allow each region to dynamically manage the situation. In other words, the creation of a single set of processes or procedures to handle such a situation would be a detriment in achieving the goal of landing all the planes. Accepting that there was no one set of processes that could handle such a crisis showed that a linear solution is not always the best solution to what would be considered a linear problem. This was an awakening for program managers as it brought to the forefront the underlying assumption that there is always one right solution or procedure to a problem inherently flawed.

According to complexity theory, humans exist together in an open system. What makes complexity theory different from a traditional open system is that complexity theory accepts that there are parts of the system that cannot be explained and that a certain degree of randomness exists (Byrne, 1998). Traditional human thought is to break down the system into its smallest parts to explain the whole. This is seen in atomic theory that attempts to explain all matter in the universe as based upon the smallest elements. Western thought seems content to understand the universe as discrete systems rather than a holistic interconnected system. This is clearly not always the case, because we cannot learn the inner workings of a colony of ants by studying a single worker ant.

Complexity Leadership Best Practice

Programs are always about people although the key deliverables might be some final product or item that did not exist previously. Because people behave in a complex non-linear fashion, complexity theory is ideally suited to apply to program management. Although the Guide to the Project Management Book of Knowledge (PMI, 2013a) and the Program Standard (PMI, 2013b) prefer a linear approach to directing a program, a leader of a mega-program must understand that his or her leadership style will have an impact on the program. The program manager needs to have the self-confidence to understand what areas of the complexity or chaos need focus and which areas need to be allowed to be resolved by those on the team. The team needs clear direction but not always detailed instructions, as micromanagement is not a successful best practice.

Seasoned program managers realize that all parts of the programs cannot be controlled, so a successful leader will need to be able to delegate in an effective and creative manner. Program managers realize that creativity occurs on the fringes of complexity or chaos, because sometimes it means giving more authority to individuals who might lack direct organizational authority. Linear thinkers can only consider evolutionary improvements where one process connects to another sequentially with the potential of incremental improvement, while complexity-based thinkers can create revolutions and introduce new systems with the potential for enormous improvements.

A single leader can motivate each individual in a program through direct and indirect actions. As explained in complexity theory, small actions and deeds can lead to large changes in a distant system, so a leader should take time for small changes to assist in creating a controlled hurricane that can achieve complex tasks. Too often people do not realize that even small contributions, such as compliments or recognition of a job well done, can build to create something greater for the mega-program. A leader who can motivate and offer small praises that matter can help move a program forward faster. The more that a program manager, particularly one that operates virtually, can harness this kind of organization, the more effective the program will become.

As a means to become a better Transformational leader, consider taking the self-assessment presented in Figure 5.1 for future improvement.

Name: Date:

TRANSFORMATIONAL LEADERSHIP SELF ASSESSMENT

Instructions	Expect to spend about 15-30 minutes with this self assessment. Try not to over analyze the question and answer with what first comes to mind. There are no right or wrong answers.

STEP 1
Do you consider yourself a transformational leader If yes, list some examples and areas to improve. If no, reflect upon what is holding you back

▼

STEP 2
Reflect upon how you can become a better transformational leader. List areas that you can take action with now.

▼

STEP 3
Consider how you can help change others to become more transformational in their leadership. In particular, consider how your changes will impact others to change

▼

STEP 4
Set a timeline for these changes in yourself and make a timeline for others to improve as well. Consider posting the plan and getting others to see it as well.

▼

STEP 5
Design a future state of how you want others to see you as a transformational leader. Offer to show this plans to others to help get them to help.

FIGURE 5.1

Transformational leader self-assessment.

Complexity Theory Showing Results Best Practice

Unfortunately, programs fail. Programs and projects, whether they are virtual or traditional, large or small, become troubled for various reasons. One example of how complexity can be applied to a troubled program is when the program manager finds himself or herself in a situation where he or she must show results before the program has attained necessary milestones. An impatient client might form a negative impression due to the lack of results. An experienced program manager will agree that stakeholder impatience and haste can often create the necessity for non-sequential activities. A linear program manager might find himself or herself paralyzed by this need which may result in the program manager pushing back upon the customer. The program manager may offer excuses or explanations which do not help the program, but may represent reality. This may cause the customer to express concern about the program, even when there is no real cause for alarm.

At this time, complexity can assist by offering the program manager a more value-driven perspective than a milestone-driven (linear) perspective. Program managers can be pushed to resolve and handle issues out of the typical sequence in order to achieve certain milestones which are important at a higher level (Weaver, 2007). This kind of pressure can be exerted upon a program manager in order to achieve certain milestones faster in order to achieve quicker results. This means that the program team must explain how the completion of later tasks can mean time saved in the future. It could also mean that a new process or system is being implemented that will accelerate the speed of completion in the future. Innovative ideas can keep a program on track, and those kinds of innovations might not appear on a static Gantt chart.

Complexity is the manifestation of empowering and delegating tasks to allow individuality to support the hive. A program manager must know that delegating groups of task can lead to synergistic creativity rather than emphasizing linear progress. This kind of thinking permeates human culture, and even television ads of insurance companies focus on bundling or combining different types of insurances for reduced rates while statistical risk analysis indicates that there is no correlation between different kinds of risks.

Complexity Theory Summary

Complexity is applicable in all areas of program management; however, applying these best practices in areas that are ambiguous can give a program manager a new tool that can be used to address these kinds of situations. Given that change is not always totally clear, the program manager would be better served to provide high-level goals and visions to better drive the leaders charged with organizational tasks. Rather than the program manager being personally involved with every change, the program manager should allow the organizational change team leaders to coalesce and come up with creative solutions to any given schedule challenge.

These best practices work best with the concept of a *federation*. As discussed before, a *federation* can assist with allowing everyone a voice in the program while still allowing for people to develop creative solutions. The program manager should ensure the schedule allows enough time and consequently enough flexibility for this important aspect of a program. Too many in leadership do not pay enough heed to the changing schedule of a program. So, a program manager must take care to monitor the changing schedule, but the program manager should learn to keep a distance to avoid becoming embroiled in the minutiae of change.

PHOTOGRAPH 5.1
Just as knowledge has grown, the castle on the hill has grown.

6

Complexity and Program Management

PHOTOGRAPH 6.0
Chaos of birds.

Just as birds about to roost at sunset appear to be a mass of chaos, the same can be seen initially of leadership and complexity. People should observe the openness and rhythm of the birds starting off as chaos and ending up in a new order.

LEADERSHIP AND COMPLEXITY

It may be said that leadership is simple while complexity is difficult; both are difficult, and leadership just adds to the complexity of a program (Figure 6.1). When discussing leadership there are as many definitions as

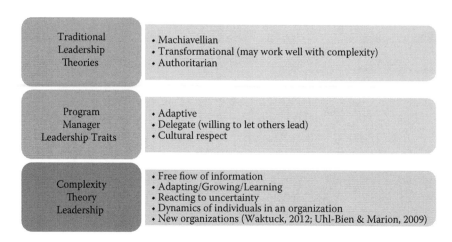

FIGURE 6.1

Leadership and the program manager.

there are individuals in the room. Academics cannot agree among themselves, and certainly, practitioners bring as many different flavors. Those who have served in the military learn early in their training that people are led and things are managed. Others believe that management is a part of leadership. This book is not to debate who is right or wrong but to offer just another hint at the complexity of humans in trying to articulate how we lead and what is better in the world of programs.

Many leadership and management theories exist. Complexity leadership theory is one of the emerging theories complementing program management. Complexity leadership theory studies the different aspects of leadership surrounding complex adaptive systems (CASs) (Uhl-Bien & Marion, 2009). Complex adaptive systems can be described in academic terms, but when it comes down to it, human beings fit nicely. Humans generally solve problems creatively, learn from situations/mistakes, and adapt (Lichtenstein, Uhl-Bien, Marion, Seers, Orton, & Schreiber, 2006; Uhl-Bien & Marion, 2009).

Waltuck (2012) studied three types of leadership within a complex environment, reviewing the type of control and the success. The styles were dogmatic, innovative (allowing for creation), and anarchistic (virtually no leadership). Waltuck (2012) plotted social data on statistical control charts. Team members on complex programs with dogmatic leaders were the least creative. Humans do not react in a linear fashion (Curlee & Gordon, 2010). To be creative in a complex, dynamic program, a change agent leader comfortable on the brink of chaos provides just enough guidance for the emergence of new ideas, adaptability, and growth (Waltuck, 2012). Being in chaos consistently showed that opinions and information were exchanged but no one would yield to compromise (Waltuck, 2012).

Program managers need to understand the art of influencing which is a soft skill of communication. An overarching leadership theme in the U.S. military is to be resourceful, to influence, to motivate, and to provide direction and purpose to ultimately accomplish the stated mission (Department of the Army, 2006). Not too long ago a leader's power was somewhat limited by physical proximity. Leaders not physically present lacked the ability to exert command and control within organizations. While these organizations do exist today, there is a transition to virtual leadership, which normally is an essential component to most programs. Uhl-Bien and Marion (2009) and Lichtenstein, Uhl-Bien, Marion, Seers, Orton, and Schreiber's (2006) studies of Complexity leadership theory support the philosophy of driving leadership to the lowest level and allowing each person to own his or her leadership. In turn, this allows the *formal* leaders to "identify strategic opportunities, developing unique alliances, and bridging gaps across the organizational hierarchy" (Lichtenstein et al., 2006).

The program manager provides strategic guidance, develops alliances (internally and externally), and fills a gap across the organization for the program. The project managers, team leads, functional managers, contract managers, purchasing managers, and others are the leaders within the program. Some have formal leadership while others will take advantage of complex situations to take informal leadership. The program manager needs to prepare each individual with the skills to take a leadership role and also the team to help the emergent leader. This is one of the main concepts of Complexity leadership theory (Lichtenstein et al., 2006).

Complexity leadership theory would advocate a structure similar to commando groups or in a negative aspect, terrorist cells. Both the commando groups and the terrorist cells are small autonomous groups which enact great change without daily direct involvement with a leader (Marcinko & Weisman, 1997). Program managers should take note of the success of these teams. These micro-organizations' leaders imbue the teams with a common culture that helps direct them toward a mutual goal. The teams come and go, but the culture and goals are imparted to the commandos or the terrorist cells with precision.

Complexity leadership theory is similar to the autonomy of a commando team. Program managers who can master injecting a micro-culture into the program become the ambassador of the culture. As the ambassador, the program manager then allows the project teams and other components to develop their own path toward program success (Curlee & Gordon, 2010; Marcinko & Weisman, 1997; Schein, 2004). Technology has increased the power of communication within the commando teams and with their leaders.

Program leadership and effectiveness have been limited by distance. Technology has helped to make the globe one large village. Leadership has failed to realize that there is still an issue with a common language. Companies may "edict" that English must be spoken; this inhibits individuals communicating in their native tongue which leads to miscommunication. Program managers, as leaders, need to have plans in place to mitigate these situations until real-time, global translations are available to each individual.

Leaders, managers, and followers all require communication. Without effective communication, any one of these elements becomes ineffectual or may even cease to exist. Leaders need to be able to communicate effectively for the situation. Many times the power relationship is about the communication relationship. The program manager must understand that he/she is limited by the communication technology. Face-to-face communication is normally the best, unless there is a language barrier. Of course, this barrier increases as technology is introduced as there are no visual clues by the speakers and no body language by the receiver which can account for up to 80% of the message. When there are communication barriers, and this will occur in a complex environment, the program manager must ensure that "barriers to communication are marginalized, and the redundancy of the communication is increased" (Curlee & Gordon, 2010, p. 111).

Technology is in transition and is changing and evolving programs, leadership, and program managers. Technology is increasingly simulating physical proximity. Once communication technology replicates physical proximity and those organizations and leaders that adapt to the new functionality harness this new power, a new dynamic can be created between leaders (program managers) and followers (program teams). Chance does not play into this new dynamic. The program manager must understand complexity theory, communication, and implementation of programs. The program manager needs to integrate the three to have confidence in his or her ability as a leader (Curlee & Gordon, 2010). Chaos may only be embraced at this point by the program manager. At this point, the program manager sees complexity taking place but is not taking advantage of its power.

Both in academia and in the practical world, individuals have come to realize that one type of leadership does not fit all occasions. This is certainly true for programs, especially the various parts of programs. Not all parts of programs are complex, and hence not all parts need adaptive or complexity theory types of leadership. Program managers may need to revert to directive leadership for those that are junior and need more hands-on leadership. For others, it may be a more mentorship-type leadership. There will be day-to-day work on a program. Do not fall into the trap of over-leading on those activities. Let individuals learn leadership on activities that are relatively safe.

Transformational leadership is another leadership theory that needs to be explored within the realm of complexity and program management. In the past, it was thought that there was one way to lead or manage people. Modern thought has radically changed. Transformational leadership is one of the modern theories that reflect society, culture, and business. Transformational leadership is a system whereby leaders and followers help each other reach higher levels of motivation. Bass and Riggio (2006) would further add an element of moral character. These characteristics of mentorship and learning align well with complexity.

A complexity sub-theory program managers should understand and embrace is the *butterfly effect*. The butterfly effect in a non-linear form is the understanding that all forces are connected (Curlee & Gordon, 2010). The program manager has the unique dilemma of meeting strategic benefits which may delay project components but yet still having to motivate these same project managers. Therefore, the program manager must communicate what is the best way to move the program forward. This involves rallying and emphasizing that all forces need to work in the same direction. This message is communicated to senior management as well as to functional management and the components. The program manager should think of the butterfly flapping its wings in Argentina and having a small effect. Over time and distance, this minute disturbance may cause a mighty hurricane in the Atlantic. The program manager's small motivations may create a tornado or hurricane within an individual or team to overcome a complex task or achieve a complex task. How many times have we seen people give up because they feel their contribution is not good or large enough? Or we have seen those small contributions initially be ignored and in the end they were the impetus to the solution. The program manager needs to positively feed the chaos to enhance creative solutions.

Change is inevitable in a program, and the program manager must connect often with governance, component, and functional leads. Complexity theory is not just a leader waiting for the chaos to happen and then issuing some orders; it is creating purpose and training individuals to take over the chaos and solidify its purpose (Curlee & Gordon, 2010). Without instilling confidence, a cadre of leaders within the team, and a definition of purpose, complexity will go unrecognized or if recognized will be ignored. Without the cadre of leaders and definition of purpose, the program manager potentially misses the opportunity for creativity, increasing revenue, decreasing costs, increasing morale, etc.

By their sheer size, many programs are virtual, where at least a part of the program is not co-located with the rest of the team. This may be a major subcontractor/vendor, one of the project components, a functional team, or the program or governance team. Virtual program leadership is different than traditional leadership. Good communications practices are a must in a virtual environment, as in most cases there is a lack of visual cues and there may be a language barrier.

There is anecdotal evidence to suggest that those program managers who incorporate a *virtual communication plan* within the work breakdown structure (WBS) and schedule have succeeded more often. These tasks address the extra needs for communication, the need to standardize technology, and the need to energize the team.

Virtual programs are about communication and leadership. For success, these two elements are necessary. The project teams, project managers, and program teams must be trained in virtual leadership, and ultimately the program manager must enforce it (Curlee & Gordon, 2010; Jarvenpaa & Leidner, 1999; Maznevski & Chudoba, 2000). Virtual leadership is based on communication and trust. The program manager adapts rules and regulations to increase the relationships and trust among the members, and between leader and member. Culture is a driving force as well. The program manager is wise to respect the culture(s) as much as possible for the diverse areas that may be represented. When respected, most team members will be motivated to provide a positive contribution.

Studies (Jarvenpaa & Leidner, 1999; Maznevski & Chudoba, 2000; Saynisch, 2010) show trust is an essential component to a virtual team; without it the team is likely to fail. Program managers who have earned the program team's trust normally succeed. The program manager's leadership style needs to establish a means to promote trust and collaboration in an environment with no body clues and many cultures. The program manager must be adept at building trust in challenging circumstances.

Lack of visual cues makes conflict more likely. The program manager must emphasize to the component leaders the need for effective communication, especially in the realm of technology. The program manager should ensure the communication plan accounts for language barriers, a common language, technology, an acronym dictionary, the need to be clear in emails and why, and other issues about *virtuality*. This needs to be driven down to the components. A training session on communication and trust, self-paced and self-taught, should be encouraged. Remember, to date, technology is not a replacement for poor communication. Also, the latest communication technology is not necessarily the best for the program. The program manager must understand the needs of the program, the geographic locations, and the capabilities of all participants. All must be included in the communication. If one component is left out, then the trust is broken. Leadership must be willing to support the virtual aspect of the program.

PROGRAM MANAGEMENT LEADERSHIP

Programs are larger organizations that involve many people in many locations. Few programs are limited to a single geographic area, and so a program leader must be comfortable working in the virtual environment. Given this, program managers should have some virtual experience in a leadership role in order to achieve long-term success. As a complex leader, a program manager must accept the fact that there can always be multiple solutions to any given problem. All of these solutions are not all equal, and so a program manager must be able to lead as well as to evaluate the different alternatives in a fashion that makes the best sense for the organization. The leader cannot lose sight that a program must meet the goals of the organization while reducing costs.

To this end, a program manager must learn to lead while leveraging technology. Since programs will often be dispersed throughout the world along with operating in different time zones, the leader cannot hope to micro-manage a program as to be omnipresent, as it was possible in the past with co-located programs. Technology can empower a leader to better monitor a greater number of people. Leaders must be willing to accept their role as farmers of their programs, because leaders of dispersed groups are impacted by natural events and the market forces in much the same manner as farmers.

The new leaders of programs must understand that the success of the group is the success of the individual. As groups grow into more social tribes, leaders must understand that they must leverage technology in order to address the needs of the program. Leaders must become more versed in directing the movement of their organization by working together within the environment rather than trying to reshape the environment. Thus, leaders must learn the importance of being in harmony with the environment rather than being in continual struggle against the elements.

Programs normally have large teams that involve many people in diverse locations. Few programs are limited to a single geographic area, and so a program manager must be comfortable working in the virtual environment. Given this, program managers should have some virtual experience in a leadership role in order to achieve long-term success. As a complex leader, a program manager must accept the fact that there can be multiple solutions to any given problem. All of these solutions are not equal, and so a program manager must be able to lead as well as to evaluate the different alternatives in a fashion that makes the best sense for the organization. The program leader cannot lose sight that a program must meet the strategic goals of the organization while realizing benefits.

A program manager needs to lead while leveraging technology. Since programs will often be dispersed throughout the world along with operating in different time zones, the leader cannot hope to micro-manage a program and be present for all on the program, as was possible in the past with co-located programs. Technology can empower a leader to better monitor a greater number of people.

The new program manager understands that the success of the team lies in the success of the individual, and as the group grows into more social entities, the leader must understand how to leverage technology in order to address the needs of the program. Leaders must become more versed in directing the movement of the organization by working together within the environment rather than trying to reshape the environment. Program managers learn the importance of the harmony of the environment rather than continually struggling against the elements.

Program managers operating in the virtual environment constantly operate in complexity. Many social systems may be explained by complexity without limiting the explanation to linear logic. Often leaders and team members demand an explanation for happenings within the program to duplicate it on future programs. Complexity provides for diverse solutions to a problem rather than the conventional single variable understanding. As organizations continue to become leaner, the program manager has to be flexible and adapt. The program manager cannot be omnipresent, as it was possible in the past. Technology has helped the leader, but the leader is no longer an engineer of people. The program manager should be a mentor to his or her team.

Teams within a program are similar to social tribes. The larger the program or the more cultures represented, the more social tribes one may expect. The program manager needs to understand that social tribes cannot be led in a traditional manner. The program manager needs to be adept in helping individuals understand the nuances of the organization's environment and how it affects the program. This affects the complexity of the program. The more tribes or teams inherent within the program will drive the different types of culture which in turn increases complexity and more fringes of chaos.

These program managers need to understand that they are not leading just a collection of individuals; they are leading various teams toward a common purpose. A program can become much more than its component parts if complexity is harnessed properly. Leaders must understand that teams are not just groups of people that happen to work together on the same program. A true team is a community of individuals that are driven toward success, and the leader constantly encourages and drives the team.

In order for the leader to start leading teams to their greatest potential, the leader first needs to "forget" what he/she knows about leadership and consider that no one leader can cover all aspects of the program. Think about an aspect of human knowledge. All the knowledge could not be concentrated into a book or taught in one course. We as individuals, especially as leaders, need to embrace that we are constantly learning and accept that no program manager can ever hope to have all the answers to all the programs on which he/she participates. The best a leader should hope for is answers to some dilemmas, good recommendations for some situations, and offering brainstorming for others. Leaders need to paint a compelling image of the new future that makes people want to travel to the new destination.

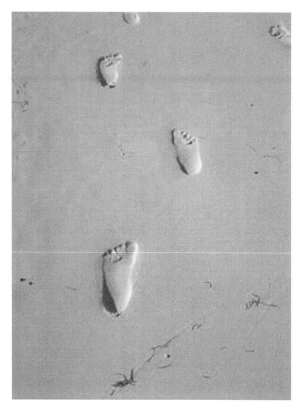

PHOTOGRAPH 6.1
Footprints in the sand: traveling from the known to the unknown.

In order to even consider utilizing complexity, one must inventory the team and take a hard look at the team dynamics. If the group is just about the program and they are only together for that purpose, then it is unlikely that complexity or any other team process will garner the type of productivity that program managers crave. The team must be motivated toward success. Team greatness is more than just spreadsheets, timelines, and milestones. Consider if the team operates as a family, as a neighborhood, or as a tribe. As a program manager, one must consider if the team is about working together and being a part of a greater whole.

This team camaraderie is often easy to spot as one can tell when individuals are actively engaged and the people are passionate about the program. Team members must enjoy what they are doing. They must see the future of the program and want to be a part of it, regardless of the outcome.

Program managers who communicate and collaborate instead of forcing agendas and issuing orders have been shown to be more successful.

The program manager must help the team reach a new level by collaborating with the program team to achieve mutual success. Being part of a successful program is good for one's career; however, individual success needs to be achieved as well. The symbiotic relationship is important and, at times difficult, since projects within the program may be delayed or terminated for the benefit of the program. The program manager needs to understand that individuals from the component project have individual fulfillment and individual pride as well. They want to be part of the greater effort, but when their objective has been undermined it is hard to keep focused. When situations of this sort occur, redirect the individuals to other parts of the program where feasible. These individuals have knowledge of the program that is most likely useful on another project component or within the program.

How does the program manager know how the team is performing? Metrics should be established. As noted previously, programs can be expected to be virtual. The metrics should demand high expectations to overcome the communications hurdles. High expectations and performance metrics should also have the secondary effect of diminishing any negative conflict. Metrics should be tied to important milestones and should be tied to incentives and celebrations. The incentives and celebrations do not have to be large. It might be recognition of the team in a newsletter, taking the team to lunch, enjoying a celebratory cake, awarding T-shirts for a larger milestone, etc. The incentives and celebrations will depend on the budget of the program and sometimes on laws and legislations. Program managers should always try to determine a manner in which to recognize individuals and teams that are meeting or exceeding metrics. Without it, the program manager will be considered a taskmaster rather than a leader.

Stretch goals are the high expectations metrics for the team and individuals. Stretch goals are the goals that are exceptional. On a program, metrics may be dynamic depending on the complexity, the strategic goals, the interdependencies, and the external influences. The program manager may have to work with each component to develop metrics that are meaningful and stable. On a program, it may not be advantageous to demand early completion for each project component. The program manager may delay a project in favor of another component to enhance the program's benefit realization. The delayed project metric cannot be held against the project component team.

A review of the Project Management Institute's Program Standard (PMI, 2013b) recognizes the importance of leadership of the program manager. Throughout the standard, almost exclusively when *program manager* was stated some form of the word *leader* was used within the sentence or at least within the paragraph. The Program Standard (PMI, 2013b) understands the program manager must be able to demonstrate leadership *up* and *down* the chain of command.

Communicating to the program strategic stakeholder community demonstrates leadership in action. The program manager must have the expertise to understand the politics, dynamics, and interrelations of the stakeholders and the influence each of them has on the program. The art of communication and influence will help the program manager manage the stakeholders. Depending on how well the program manager has researched each stakeholder's influences, needs and relationships may very well determine how successful he/she is with a successful negotiation with any one stakeholder. For example, the program manager has done some analysis and realizes that slowing down project A and accelerating project B will increase benefit realization for the program twofold. Both of the project sponsors sit on the governance board for the program. The stakeholder for project A needs to be in favor of slowing down his/her project for this suggestion to be successful. A tactic to achieve stakeholder A's approval would be to speak to the stakeholder prior to the governance board's meeting and demonstrate to the stakeholder the benefits to the program and the company. Following the meeting with the stakeholder, a meeting should be held with the program's sponsor so there are no surprises at the governance board meeting.

PHOTOGRAPH 6.2
Finding the new destination.

7

Course Materials #2

PHOTOGRAPH 7.0
Can you build a successful organization without having a sound foundation?

———————

CLASSROOM MATERIALS

Case Study #2—Building and Leading Successful Organizations

You have been hired as a new vice president in charge of the Drone Program for a growing midsized company that specializes in the manufacturing of military electronics. The company has a number of issues, but the owners have decided to bring in a new person to take on the task of heading up a new program that is to design, build, service, and maintain the next generation of drone technology. This next generation of drones will have limited artificial intelligence capabilities to operate independently. This will allow the drone to operate without the need to remain in contact with the soldiers in the field so that these drones can avoid being disrupted by either battle damage or enemy infiltration. The owners now recognize that the competition is improving, and so they have brought you in to head up this high-profile program. Their hope is that you will be able to improve productivity in order to enact sweeping changes throughout the organization.

Company Details

The company is called Sterling Defense Systems, and it is a manufacturing company of military-grade electronics and guidance systems. Sterling specializes in high-end, highly reliable, and cutting-edge-type items. The company has a 25-year tradition of producing reliable items. Sterling does some retail-type electronics business, but the primary business has been with military work both in the United States and internationally.

The company is based in Los Angeles, and there are also offices in Rotterdam and Rome. The Los Angeles office handles sales, distribution, as well as customer service. The Los Angeles facility houses manufacturing, research and development, marketing, supply chain, distribution, and reverse logistics. All manufacturing is done in the United States, but many components are purchased overseas. The offices in Rotterdam and Rome are primarily for sales, but they also do some logistics for European parts.

This program will have three major project implementations: the design project, the build project, and the service and maintain project. The company has a number of different people available to head up each of these component projects, all of whom will report to you.

Summary of Employees

Kate Coffee

Kate has been with the company for 16 years and has been the project head of several successful projects. She is well liked and well regarded in the organization. She is considered to be organized and driven. She does not work well with William White because she feels that he is too focused on the numbers rather than on the needs of the customer.

Dieter Dime

Dieter has eight years with the company and is considered by many to be a good engineer. He is known for finding low-cost solutions to difficult problems. Most people can work with him, and he can be a good mentor to younger engineers. He does not work well with Terry Zach.

Andy Green

Andy has been with the company for 30 years and is looking toward retirement. Andy is looking for one more great project to end his career at Sterling. He is a good engineer and has been on many successful projects, most of them related to design. He is respected by most people in the company, and he is often consulted by others in the company due to his vast experience. He is a great leader and although some people do not agree with his style, they all respect him. He can be stuck in his ways at times and sometimes resists change. He does not work well with Dieter Dime, because he views Dieter as always looking for cheap solutions rather than ones that are built to last.

Sam Jackson

Sam has 10 years with the company and has been involved with several successful projects. He is considered a great cheerleader of successful programs and often can garner stakeholder support. He is not afraid to spend a lot of money marketing a program, and some view him as not fiscally minded. Sam is a good transformational leader and is well liked and can work with most people, but he prefers not to work with new employees as he finds them a drain as he has to spend a lot of time explaining basic company matters to them.

John James

John is new to the company and has little experience with projects. He is just out of college and was a part-time intern with the company during college. John is eager and willing to learn. John is a nephew of Susan Johnson.

Susan Johnson

Susan has eight years with the company and has been a successful project manager for a few midsized projects. She has five years of experience as a project manager, and although she has done well, she is not well versed with international projects. She has a total of 10 years of experience with projects and programs (two years with a competitive company). She is great working with clients, and she has a very positive and enthusiastic leadership style. She works well with most people and has been trying to mentor her nephew, John James.

James Kowalski

James has been with the company for six years and has no other experience as he started here right after college. He is very technology minded and has a good mind for logistics. He has done well supporting programs but has expressed interest in leading a project. James is willing to work with anyone except Tim Topper.

Tom Lee

Tom has been with the company for four years and has an additional year of experience with another company. Most of Tom's experience has been in production, and he has assisted with several successful programs. He is considered enthusiastic and organized and most people enjoy working with him. He is very gregarious and tends to socialize well with others. All of his clients have always had positive reviews about his project performance. Tom can work with anyone but prefers to work with Todd McNulty or Rich Ready.

Ralph Marigold

Ralph has eight years of project management experience and has been with the company for 10 years. Ralph is good at motivating people and knows how to get people to work hard toward deadlines. Ralph has a good reputation with the company and is seen as a person who will help with any project that needs help. He tends to be moody, which puts some people off, but he has solid skills and has had a few projects that were so successful that people still talk about them. Ralph is a laid-back leader who does a good job delegating. Ralph likes to work with William White as he feels that William always watches the numbers and helps make the project very successful.

Todd McNulty

Todd has 15 years of experience with programs and has led many successful projects. His last project was cancelled after budget overruns in the engineering phase, but most people felt that he did a good job. Todd can be very demanding, and he tends to make people work hard at all times. Most people find his leadership style to be straightforward. He likes working with Dieter Dime.

Pat Nathanson

Pat has two years of experience with projects; however, his last project was cancelled. Everyone said that he did a good job with that last project, and they felt that Pat was part of the reason that the project was not cancelled sooner because the client felt that Pat did all the right things to keep the project on track. Pat is a relatively untested leader, but most believe that he will do great things in the future.

Tim Topper

Tim has seven years of project management experience, five of those with the company. Tim is seen by many as unlucky because most of Tim's projects have been cancelled before they are completed. Due to this, some call him the cancel man and do not like to work on projects with him. Tim is good with timelines and organization and has been known to do a fair job with numbers. He is a very motivated individual and is always looking to push his team to do more. Tim found working with Pat Nathanson a good fit, because they have very similar leadership styles.

Rich Ready

Rich has 12 years of experience and has been with the company for 10 of those years. He is considered to be a detailed and driven individual who is very safety conscious. He works well within a team environment and always keeps safety in mind. He is very social and can help move projects along safely. He has headed up a few successful projects but does not mind working with others who have more experience. Rich prefers not to work with Tim Topper, due to some personality clashes during some heated project meetings in the past.

George Samson

George is right out of college and people have high expectations of him. He joined the company just a few months ago and was recommended to work at the company by Terry Zach. He graduated at the top of his class, and people think that he will move quickly in the company. People already find him to be very organized and bright, and he tends to pick up technology very quickly. He is interested in doing more in engineering as that was his field of study in school.

Shannon Smart

Shannon has four years of project management experience and is good with creative finance. She is very focused on the financial health of a project but can find new ways to squeeze out good value for clients. She is a tough negotiator, and people find her very good with following up and with meeting deadlines. She prefers not to work with Kate Coffee, as she felt that Kate did not reward her as much as she deserved in the last successful project.

Kathie Smith

Kathie is a new employee to the company, but she has 10 years of program management experience. She is considered strong willed and determined, and although people feel that she will contribute in the future, she is untested with others in the company. She is a great transformational leader and in the past has been known to turn around failing projects. She is a good manager of people, and she is very keen to begin her first challenge with the company.

Michelle Thomson

Michelle has been with the company for a year and is seen as very enthusiastic and very interested to work on any project. She is seen to be a real go-getter and is very social and good with people. She has made some mistakes on some important documents in the past, so many do not see her as good with those types of details. She is eager and people find that she has a creative mind and can often find creative solutions to difficult problems.

William White

William has 10 years with the company and is very financially driven. He is good with numbers and is seen as an asset by many due to his ability to handle accounting problems. He is a professional and organized project manager but tends to be more of a financially driven leader. He can be socially shy at times and does not always do well with very social clients.

Terry Zach

Terry has 20 years with the company with all of those years heading up projects and programs. Most of his projects have been successful. For the few that were cancelled or not successful, he still managed to earn the respect of others in the company and of the clients. He is feared by some because he is very politically minded and will act in his own best interest at times. He has been known to blame others at times, which makes him an average leader. He has a good mind for numbers and also keeps a strong watch with regard to the financial health of a project.

Assignment

1. Given the three projects that are needed in this program, identify the project manager and an alternate for each project.
2. Justify your decision and explain your rationale for each appointment.
3. Identify at least two others who should be part of each team.
4. Justify your decision and explain your rationale for each team.

SECTION QUIZ

This section is separated into three parts. Section 1 has multiple choice questions, Section 2 has true/false questions, and Section 3 provides answers to the questions in Sections 1 and 2.

Section 1

Multiple Choice

1. What are the four eras of leadership?
   a. Early, Industrial, Modern, Advanced
   b. Early, Late, Industrial, Modern
   c. Early, Industrial, Modern, Post-Modern
   d. Transformational, Early, Late, Trait

2. What four leadership theories have dominated throughout time?
    a. Great Man, Trait, Type A, Contingency
    b. Great Man, Trait, Contingency, Transformational
    c. Trait, Contingency, Transformational, Stewardship
    d. None of the above

3. Which leadership theory works best with complexity?
    a. Trait
    b. Contingency
    c. Linear
    d. Transformational

4. What best practice can assist in everyone having a voice within a program?
    a. Non-linear thinking
    b. Federation
    c. Contingency
    d. None of the above

Section 2

True/False

1. A single leader has the potential to motivate everyone in a program.
    a. True
    b. False

2. Program managers should not use a schedule because it is too linear.
    a. True
    b. False

3. A good program manager will be in control of all aspects of a program at all times.
    a. True
    b. False

Section 3

Answer Key

Section 1

   1. C
   2. B
   3. D
   4. B

Section 2

   1. True
   2. False
   3. False

DISCUSSION QUESTIONS

Short Discussion Questions

What is leadership theory based upon?

What type of leadership works well with complexity theory?

Explain the different eras of leadership.

Explain which era of leadership is most beneficial to you.

Explain which era of leadership your organization operates from.

Give an example and explain a successful leader for a program or project.

Give an example and explain an unsuccessful leader of a program or project.

Explain how technology can assist leaders with program management.

Give some examples of complexity theory being applied by leaders within a program.

Long Discussion Questions

Define and discuss the evolution of leadership theory. Give examples of each and indicate the significant impacts of the history of both. Explain how and why modern leadership theory is important in program management.

Give an example of a successful program that was able to leverage transformational leadership. Explain the importance of the human factors in the deployment of transformational leadership within an organization. Support your position.

Research complexity theory and transformation leadership and offer recommendations on how both could have been deployed successfully in a previous program. Defend your answer and use support to show how this approach would have been more successful than another methodology.

Define program management. Offer a potentially successful linear method to lead a program, and then offer a potentially successful non-linear method to lead a program. Compare and contrast the two positions and offer your recommendations on both.

SUMMARY

Seasoned program managers realize that a complex program has so many parts that it is impossible for even the most detailed micro-manager to control it (Jaafari, 2003). As projects have become larger and more diverse, programs have evolved to being even more diverse and complicated. A single manager does not have the time in a day to manage every aspect and element of a program. This has forced individuals to realize that there are limits to control; however, they must then deploy new methods to organize the program. The program manager must become more creative in monitoring the fringes of complexity or chaos.

Some program managers who have coped with *edge of chaos* situations have often leveraged conventional linear thinking with practices such as delegation, empowerment, and stewardship. The difference is that these named management theories are about taking action or controlling singular actions or tasks. Employees can be delegated authority to handle tasks, employees can be empowered to make certain decisions, and a leader might position himself or herself as a steward of the organization; however, all of this falls short of complexity. These theories and management ideas are important and effective; however, the difference between complexity and these others is the role of the manager. In all of these prior theories, the leader was always the final authority in the paradigm. The program manager was the only authority, and the program manager set the rules and his or her rules were the law (Bass, 1990).

A linear program manager is one who followed the example of the leader of an artillery team. In a linear organization, artillery fire was handled by a team of individuals who were focused on loading and firing the artillery. The artillery team was directed by a single individual who would monitor and gauge the effectiveness and accuracy of the projectile. The forward observer would then direct the team in order to modify the direction of the fire. The forward observer's decision was then executed by the artillery team (U.S. Army & Marine Corps, 2007). This is the template for a classic linear program. In this case, the program manager assumes the role of the forward observer, and the rest of the project team was responsible for loading, firing, and aiming the artillery.

Military equipment has advanced to the point where targets are identified and smart missiles fire upon a target. There is no longer leader interaction to direct this kind of fire. This fire-and-forget technology is akin to complexity theory. Targets are identified by technology regarding which are friend or foe, and then the individual, assisted by technology, fires upon the largest threats or targets where they are most likely to eliminate. Program managers must learn to turn loose their teams upon problems/tasks/duties and then allow them to solve them and move on (U.S. Army & Marine Corps, 2007). The overriding assumption is that individual projects within the program are completed and not left incomplete. Leaders need not follow up with others on the team to know if assigned tasks are done. Assigned tasks are completed on time, and the team moves on to the next task. This is the new paradigm that embraces complexity.

What disturbs most people is that this kind of thinking makes a team appear to be in total chaos. Since individuals are expected to do their best work for the program, some work teams may experience a catastrophic failure of some sort; however, they are allowed to resolve it on their own. Making the team handle its own failures will force the team to accomplish this task quickly and efficiently (Hass, 2009). This appears to go contrary to standards suggesting a linear solution or program managers who want to be consulted with problems to offer their expert input. The weakness with that is that problems cannot be addressed as quickly as they arise, because people feel the need to explain the options. This gives the appearance that a program is totally out of control, because teams are not alerting the program manager of potential issues. Furthermore, when there is no formal reporting structure to ensure that all matters are handled, people feel that they are being excluded; however, one must foster an environment where professionals are addressing problems and moving on to the next task without intervention (Hass, 2009).

To be successful with complexity, the program manager must completely understand the theory and implementation within the scope of program management. The program manager must also make sure that he or she has the right people in the right spots, because having a linear thinker who is always waiting for direction will be problematic on many levels. The program manager must understand the integration of the processes and must have confidence in his or her ability as a leader as well as confidence in the people to *always* do the right thing (Hass, 2009). Once the right people are in place, then the program manager is ready to embrace the ability to allow *chaos* on the project. If the program manager is not willing to let go to this level, then it is not recommended to move toward complexity.

PHOTOGRAPH 7.1
Will you build a good team or will you walk alone?

8

Communication and Program Management

PHOTOGRAPH 8.0

A foggy beach at sunrise is like communication in program management. A program manager can try his or her best to communicate, but there is always a fog of distractions that get in the way of good communication.

INTRODUCTION TO COMMUNICATION

Organizational communication is complex and important to the virtual organization. No organization is fully co-located with all of its stakeholders, customers, suppliers, and employees. Organizations are no longer housed in a single building or area, and technology is making it less likely that this will ever be the case again. Although many organizations strive for greater face-to-face communication, there are other technology-based communication methods that perform the same function as these face-to-face meetings. The continuous monitoring of team members that organizations of the past felt was essential is no longer seen as an organizational priority. Individuals involved with a program are expected to perform their tasks with others in mind so that the program can achieve the stated objectives.

Technology cannot replace poor communication; in fact, effective communication is one of the most critical elements of a virtual team. Communication is moving away from the point-to-point, face-to-face communication of the past. Face-to-face communication is being replaced by different methods of communication. Telephones and electronic mail have replaced face-to-face communication with regard to passing along stories that support organizational lore. These methods of communication are informal and more readily simulate an oral mode of communication.

The contemporary organization must learn to communicate better so that messages are received and comprehended. The communication distribution system has emerged as an important force in effective internal and external organizational communications. This is particularly important in virtual organizations because one needs to be sure that people involved in a program not only receive the necessary information but understand the current and future intent of the program.

Communication Summary

The virtual environment continues to evolve, and because of this evolution, there is great speculation about the future direction of virtual organizations. The new organizations that can adapt to change will become even more important, and this is essential for program management. Programs are not static units that will remain the same throughout the life of the program. Programs must have built-in adaptability, and they must be designed with the future in mind. Program managers who understand that technology will be changing rapidly and have a plan to be able to upgrade incrementally will do far better than those program managers who design programs for a set, fixed purpose without an eye toward the future.

PHOTOGRAPH 8.1
A virtual organization is like a spiral staircase into the sky.

Virtual Organization

A virtual organization is a group of individuals who work across space, time, and organizational boundaries. This group is brought together to perform interdependent tasks, while occupying geographically different locations which are united by communications technology and a common purpose. Or, more succinctly, a virtual team is a boundaryless networked organization assembled to perform a task where the team is coordinated through trust and shared communications. The virtual environment combines elements of virtual teams and virtual organizations; hence, the virtual environment is the boundaryless, networked work setting that binds geographically distributed individuals by communications technology.

Economic pressures and the desire to hire the best talent drive companies to use virtual organizations. The need for flexibility within the organization also drives the need for virtual teams and virtual organizations. It is feasible that in our lifetime, virtual organizations may become more commonplace than brick-and-mortar organizations. In the future, organizations may have to justify the need to bind their organizations and teams to one location when it might be expected to deliver the same function, with fewer resources, at a lower overhead, and with higher morale from a virtual organization.

Organizations should expect to have virtuality be a part of the business landscape. Virtuality needs to be embraced. Leaders need to start thinking in this manner and applying technology to solve the problems of virtual teams. The technology cannot just be adapted from traditional brick-and-mortar and made to work in the virtual environment. The social environment is constantly moving—one only has to look at a Facebook* account. Teenagers have taken social media to another level of communication. Teenagers and young adults can conduct an entire day of business without ever verbally communicating. They effectively use instant messenger, email, texts, tweets, and other social media. Personal and business lives, if not careful, can blend into one. Leaders of virtual organizations need to harness social media to creatively capture the chaos of business in the virtual environment.

Today, a virtual organization could be described as more than 50% of the team members not resident in the same physical location but not necessarily dispersed over different time zones. The team depends on technology to communicate, rarely or never meets face to face, and team members make decisions among themselves (Kelley, 2001; Maznevski & Chudoba, 2000; Townsend & DeMarie, 1998). The team members rely on technology to communicate with each other, the customer, the leader, and other external parties to accomplish the organization's objectives (Reinsch, 1999). Virtual teams or organizations may also be known as telecommuting, distributed, disbursed, among others.

Before continuing the discussion of virtual organizations of today, let's discuss the history of virtual organizations. Virtual organizations were seen with Moses in the Bible, the Romans, and the British Empire. In the Bible, Jethro chided Moses for not delegating daily tasks to able men (Exodus 18:17–23, King James Version). Moses does as his father-in-law advises and chooses responsible men and makes them leaders throughout Israel (Exodus 18:25, King James Version). This is management by exception (Shafritz & Ott, 1996), and this continues through history. Caesar used it throughout Rome and it was used through the British Empire.

Communication extended virtually through the placement of rulers in outlying lands and their human messengers.

Virtual organizations are different today since there is use of technology and communication is almost instantaneous. Other differences are that leaders are worried about organizational budgets, diminishing risk, and meeting timetables. These differences from a project management discipline drive the organization to portfolios, programs, and projects.

Programs are risk laden by their nature, and by adding the virtual aspect they become even more risky. Risk in ancient times was not a consideration. Resources were plentiful; schedule and costs were not an issue. Program sponsors were dictators or tyrants which allowed them to take control. Today's Western business world's law and culture would prevent one person from taking absolute control (Gordon & Curlee, 2011).

There are seven virtual teams identified by Duarte and Snyder (2006). The networked and project teams identified by Duarte and Snyder (2006) are similar. The other five virtual teams do not have the commonalities to a program. The network and project (which can be extrapolated to the component projects) teams cross geographic and organizational boundaries and time, and both have a common purpose. The component projects maintain cohesiveness for a set period of time and perform tasks for the program. The networked team has a less defined life span (the program), and tasks may be more predictable and routine (the management of the program) (Gordon & Curlee, 2011).

The program manager deals with networked and project teams on a continual basis. The astute program manager enhances the communication between the component project teams and the program team (network team). This enhanced communication process helps the overall success of the program and creates an understanding for the project team when the program does odd things to the component project. The odd things may accelerate the project or slow or even stop the project. The program leaders need to communicate the "why." Networked teams and project teams have individuals come and go and both are working toward a goal. Each needs to understand why each other's goals may be in conflict.

Virtual organizations must be ready to work on the edge of chaos as complexity is more inherent in them than with traditional teams. Managing technology, geographical boundaries, time zones, and organizational boundaries are some of the difficulties facing the disbursed team. Duarte and Snyder's studies (2006) also found that collaboration of virtual teams is difficult due to differing working environments, culture clash, incompatible technologies, and incompatible goals.

Companies of all sizes with programs that have a virtual aspect should review their policies and procedures. A virtual program is not a traditional brick-and-mortar program just with some resources in a disbursed mode. Studies indicate that technology is only one factor to make a virtual organization or team successful. Duarte and Snyder succinctly encompassed the other factors in six areas: human resource policies, training and on-the-job education and development, standard organizational and team processes, organizational culture, leadership support of virtual teams, and team leader and team member competencies (2006, pp. 12–13).

Reinsch's (1999) studies of virtual organizations indicated that a strong and stable relationship with a supervisor greatly increased the success of the environment. Within a virtual program environment, this stability is not the norm. Most programs consist of a team of individuals who may or may not know each other, and who are assigned to a program in a matrix management relationship.

Roberts, Kossek, and Ozeki (1998) found that executives dealing with virtual organizations had three common issues: ensuring the correct skills were in the correct region/area when needed (¶ 13), disseminating innovative and "state of the art knowledge and practices" (¶ 14), and identifying the talent throughout the organization (¶ 15). English was the business language for all the companies within the study. However, this hindered the virtual organization because of the different English grammar, English not being a native language, and the nuances of the various English versions.

Boudreau, Loch, Robey, and Straud (1998) found that virtual organizations that leverage the notion of a federation are more successful than those that do not (¶ 13). The federation concept as defined by Boudreau et al. (1998) is virtual partnerships, joint ventures, consortia, and other alliances that are managed by a group are designed to change with a program. A federation may include alliances with other outside organizations or stakeholders involved with the success of a program (Boudreau et al., 1998). The federation concept has been applied successfully to the B-1 Bomber program, which had over 2,000 corporations working together, most of which had primary interaction that was virtual. Other successful corporations that utilize a federation concept include Sun Microsystems, Nike, and Reebok (Boudreau et al., 1998). A federation can help build an organization that is focused upon the success of a program. A federation that works together will be able to work out scope details better than an organization that lacks a process that involves everyone within the scope of a program.

The seamless integration of the technology within the organization and among the federation members (Boudreau et al., 1998) allows the program to have the support of a worldwide virtual organization, and the client does not realize that the product is a result of several companies or organizations. A well-run virtual organization should be able to function with very little regard to geographical distance and time barriers. According to Boudreau et al. (1998), a well-run federated virtual environment must be technologically seamless, responsive to local needs, and have the centralization necessary for efficiency.

Technology may be the downfall for a virtual organization. When conducting a technology gap analysis for the infrastructure, the projects, programs, accounting, and other systems, leadership may fall into the trap that the latest and greatest may be the best for the company. This may not be the case. A thorough analysis of the company's industry, where the company does business (such as in developing countries), what type of laws and legislation have to be embedded in the systems (such as Sarbanes-Oxley or host country laws), and how the technologies have to integrate with other companies or partners on projects or programs or for procurements or with clients, needs to be thoroughly assessed.

When dealing with developing countries, those making decisions must understand that the infrastructure may not be stable. The Internet may not be able to handle streaming video. There may not be steady voice availability. Even dealing with various companies within one country may present an issue with a mundane thing such as email. Consider Lotus Notes and MS Outlook. The two are not compatible in many areas and even email messages can be lost. This technology has caused many an issue for many organizations working with each other because the incompatibility was not addressed.

Organizations today are normally hybrid. Parts of the organization are virtual and parts are traditional brick and mortar. Within the same day a person could be a virtual employee and a brick-and-mortar employee. The leaders of the organization need to have policies in place to help the employee not be hindered from transitioning from one role to the other. For example, consider local travel between sites, is it reimbursable? How about parking if the employee had to travel? How about the law? How does the Racketeer Influenced and Corrupt Organizations Act (RICO) or the Federal Corruptions Practices Act (FCPA) affect the employee? If employees need to know about a certain law, how do they find out about it if they are not at the traditional brick-and-mortar location? Is there training? Is it implemented in the processes/methodology/procedures? These are areas where a human resources specialist or a general counsel can help with training or guidance.

Virtual organizations are difficult for some individuals to adjust to as it can be a lonely transition. Training must be provided by the organization to help the individual cope, and tips should be provided on how to adjust. Not everyone may be ready to work without the daily interaction of his or her colleagues just an office or cubicle away. The shift to telecommuting and having to reach out via IM, social media, email, or phone may be lonely for some. These individuals may need to think hard before switching to the virtual environment. The organization may have the technology to provide face-to-face meetings which may appease the loneliness. Supervisors may need to reach out more often to those in transition. Leadership needs to assess the loneliness of its workforce and also understand that there is a learning curve to moving to the virtual environment.

In summary, virtual organizations are the new business landscape and will continue to grow. Those leaders able to harness social media into the virtual landscape will exponentially make their businesses grow as they are able to harness the edge of chaos. These visionaries understand that technology and using it in new ways will capture niches of their markets that others had never envisioned. Technology and virtual organizations are changing at such a rapid speed that academics and practitioners are finding themselves constantly behind in trying to give *sage* advice. The best advice is to brainstorm when complexity provides the edges of chaos.

Why Companies Are Going Virtual

Organizations must become chameleons in order to survive in business today. This creative metaphor offers a view of the proper relationship of a future organization. The organization of the future must morph into an adaptable organism. Its color, shape, size, and appearance will change as its environment and the demands placed on the organization evolve. This distinctive metaphor describes the elements of virtual organizations. The organization is flexible and adapts to the environment rather than making the environment adapt to the organization.

Another metaphor for the virtual organization is that of corporate condominiums. Implying that the companies of the future are condominiums is an excellent example of using a metaphor to describe this new model of business. Virtual organizations are more than just temporary brain trusts, they are organizations of greater ownership than those that just exist to work in an organization. Real ownership implies a sharing of the wealth and a sharing of the information of the organization. The sharing of knowledge can happen in a virtual organization where every person can feel that he or she is a valuable member of the whole. Virtual organizations are going to be very different from the organizations of the past.

Building Trust in the Virtual Organization

For a virtual organization to be successful, trust must be part of the foundation of the organization. Trust demands boundaries and learning. Trust requires bonding and leaders who have a certain touch to management. Trust includes such factors as competence, integrity, and a concern for others (stakeholders). These elements of trust are necessary to make an organization successful. Trust must be visible at all levels of an organization in order to achieve greatness. No virtual organization can be successful, let alone effective, without the individuals involved having a high level of trust at all levels.

There are many elements that go into trust (see Figure 8.1). One should consider all the elements mentioned in the puzzle so that they can be integrated into a program.

The program manager's leadership style benefits by promoting trust and collaboration in a faceless environment. Creative manners and communication help the program manager establish trust. The effective program manager establishes trust between himself or herself and each team member individually. A program manager must also promote trust within the group. Devoting time to building relationships with each team member is a must for the program manager. The program manager must learn how to interact with each team member and to keep track of their individual progress. A program manager must be interested not only in the program, but in the person. This will make a difference when building trust. Individuals trust those who know them.

Trust does not end with the individual; the program manager must build trust as part of a group. Group members must learn to trust each other and to trust the judgment of the group. Trust is created when individuals respect one another. Trust must be visible at all levels of an organization in order to achieve greatness. No virtual organization can be successful, let

Name: Date:

VIRTUAL COMMUNICATION SELF-ASSESSMENT

Instructions	Expect to spend about 15-30 minutes with this self assessment. Try not to over analyze the question and answer with what first comes to mind. There are no right or wrong answers.

STEP 1
Do you consider yourself a good virtual communicator? If yes, list some examples and areas to improve. If no, reflect upon what is holding you back from doing better.

▼

STEP 2
Reflect upon how you can become a better virtual communicator. List areas that you can take action with now.

▼

STEP 3
Consider how you can help change others to become better virtual communicators. In particular, consider how your changes will impact others to change

▼

STEP 4
Set a timeline for these changes in yourself and make a timeline for others to improve as well. Consider posting the plan and getting others to see it as well.

▼

STEP 5
Design a future state of how you want others to see you as a high effective virtual communicator Offer to show this plans to others to help get them to help.

FIGURE 8.1
Trust puzzle.

alone effective, without the individuals involved having a high level of trust between one another: A program manager should avoid solely using strategies that create trust for a group. This is a common error made by leaders of virtual organizations. For example, a team meeting will not build trust when people report only on their progress. A staff meeting does nothing to build trust if trust is not on the agenda. If a program manager wants to use this time to build trust, then put trust at the top of the agenda and stick to it. Talk to the team about trust and have team members talk to each other about what trust means to them. Communicate about how effective trust is (or is not) being built in the program. The results might surprise you once you have a group talking about trust. A program manager must remember that building trust is not an either-or proposition. Trust must be built individually, and it must be built as a group. Studies have clearly shown that without trust, a virtual team is more likely to fail.

One study found that executives in eight major U.S. corporations agreed that it was difficult to establish trust in a multi-cultural environment. This lack of trust led to the companies establishing duplicate processes and procedures and different systems, which results in many international companies instead of one cohesive enterprise. If supervisors do not already trust and respect virtual employees, then they need to build trust and respect for virtual employees. Many managers need visual clues and interrelationships to be comfortable. A virtual program manager needs to learn how to move past this managerial limitation and how to rely upon other cues for security.

Trust and Communication

Organizational communication is complex and important to the virtual team. Organizations are no longer visible, tangible places where managers can wander around to learn about what is happening. Management by walking around is no longer a viable option in the virtual world. Virtual workers need to learn to better communicate because there is less contact time for individuals. As technology improves and becomes less expensive to implement, virtual organizations will become the norm. People are not connected only by phone, fax, and email. The continuous monitoring of team members that organizations of the past felt was essential will no longer be necessary. Work will be done in a manner where no one can see it happen. Individuals in organizations of the future will learn to make a few bold strokes and then pass the brush along to someone else who will then add his or her own perspective to the growing work of art. To meet this new way of conducting business, a program manager will need to increase the level of communication.

There are three views to communication. In the end, all of these styles of communication can be effective. A program manager has to decide which one(s) they will use consistently and then use them. Communication builds trust when there is predictability. First, one can approach communication as evolving from the face-to-face communication necessary in hierarchical structures to a vehicle to convey culture through stories. Some managers feel that if they can tell a story that parallels the current situation with a positive outcome, they can help individuals deal with the current situation. In the past, managers would build this trust through face-to-face communication; however, that option is no longer possible.

Second, those who enhance communication and trust building can be seen as gatekeepers, individuals who control the flow of communication to others. This view addresses the virtual relationship of individuals as mediators in a dynamic communication system—a relationship of cause and effect. Gatekeepers in the virtual environment can either restrict information or mediate communication. Either form can be altruistic; however, these gatekeepers can be far more insidious as they can be skilled politicians who control knowledge to maintain their expert status. Although a manager might feel more in control when he or she uses this method, this form of restricted communication will always be seen by others as negative. When information becomes a medium of exchange, it can create some relationships that are not founded in trust. The program manager must use this kind of communication carefully.

Third, in contrast to these views about gatekeepers, others believe that the virtual organization is not about the single steward whose actions control the organization's destiny. Leading a virtual organization requires a single individual who harnesses a personal and business network in order to achieve impressive results. Virtual teams require that leaders shift from a focus on personal ability to a focus on group results. Individuals who are interested in leveraging the strengths associated with virtual work must increasingly follow this pattern of success through networking in the future.

Ultimately, organizations must learn to communicate better in order to build trust and to ensure that their messages are received and comprehended. The communication distribution system has emerged as an important force in effective internal and external organizational communications. This is particularly important in virtual teams because of the potential impact upon the virtual team.

Trust and Change in the Virtual Organization

The important reason for building trust in any organization is the need for trust in order to enact change. Change requires planning, organizing, controlling, and leading. Planned change never happens in a vacuum. Managers who want to implement change must learn to lead. Leading for change takes a good leader and good followers. The leader must not only know his or her followers, but a leader must trust them and vice versa. So how does this lead to change?

Effective change comes when leadership plans it; however, the largest obstacle to change is always resistance to new ideas. A leader must have the respect of his or her followers and must remain steadfast whenever encountering resistance. Anyone can steer a ship when the seas are calm, but it takes a true leader to maintain a steady course when the seas are rough. To implement change, one must learn to lead. To overcome resistance to change, the leader must understand certain realities associated with any change. The reality is that the manager must modify his or her style to accommodate the situation. There is no one right form of leadership, just as there is no one right organization. The more techniques a person can learn will offer that person greater flexibility in meeting the ultimate goal of leading change. Review the flow of change in Figure 8.2 in order to improve trust within an organization.

Leadership, Competency, Technology

Leaders must support the team as well as help them move toward the desired goal. However, task leadership was not sufficient for success in the virtual environment. Leaders who only communicated tasks and timelines were not as successful as those who truly led the team. Leaders not only had to help the team toward the new goal, but they had to identify and garner support of the extended network of experts who supported the group. The program manager identified these stakeholders and gained their support in order to make the program successful.

One successful strategy observed was the creation of a stakeholder matrix outlining level of participation, roles, and contact information. This was important to allow all team members to understand what resources

Virtual Program Improvement

Trust
Earn Trust
Build Trust
Grow Trust

Change
Explain the change
Review the change
Sell the change

Communicate
Use different types of communication
Make contact multiple times
Leverage different media

Improve
Improve the people
Exceed the expectations of the client
Improve the program

FIGURE 8.2
Virtual program trust improvements.

were available to the team. This not only communicated to team members the listing of stakeholders, it also helps them understand the important participants in any program. A team that becomes a tool of positive communication to stakeholders can become a powerful tool in program success. Furthermore, successful virtual program managers were skilled at gaining support of customers and stakeholders. Program leaders must be able to not only rally the troops, but they must be skilled at rallying support from all areas of the organization toward the program. The program manager must build a stakeholder matrix and communicate this information to appropriate team members. Ask team members if they ever pass along information regarding the program to these stakeholders. Creating positive marketing of a program will help ensure its success.

The successful program manager understands the needs of the members and adapted rules and regulations to increase the relationship and trust among the members and between leader and member. A best practice for a virtual program leader is to adapt the competencies of different cultures. Trying to go with the flow rather than fighting the current is a wise move for a program manager. Understanding and then leveraging each team member's strengths is critical. Team members will be motivated to contribute positively to the group, and they will be proud to have the opportunity to contribute to the group.

Other successful leadership strategies for successful programs are as follows: developing and transitioning team members, developing and adapting organizational processes to meet the team's needs, allowing leadership to transition when appropriate, and ensuring the team received appropriate training for virtual communications and technology and skill sets.

A program manager must consider if he or she has the ability to develop and transition team members as the program requires. All programs have a degree of transience, and this means that team members will shift in and out of the program as the various milestones are reached. Given that all team members will not be active throughout the program, a program manager must expect that team members will enter and exit the program. Being able to transition these team members rapidly and effectively will assist in making the program successful. Although this may not seem directly related to the program timeline, it will affect the overall morale, trust, and feelings of the team members.

A program manager should review the changes that will be required by the program and then map out the new organizational processes that will be able to meet these new requirements. Leadership transition is important to the success of a virtual program. The program manager will not always be the person who is best suited to lead the team at all times. There will be times when it would be better to allow someone else, perhaps with more technical ability, or perhaps an influential stakeholder, to lead the team on an interim basis.

The program manager should also establish metrics that measure team performance. These metrics should institute high expectations to encourage the extra effort required to overcome the communication hurdles (Duarte & Snyder, 2006). As the team becomes familiar with the expectations and the performance metrics, the amount of undesirable conflict should be reduced.

Strategies of Successful Program Managers

From a host of potential strategies available in the current literature, the experts most commonly recommended four winning strategies. The virtual program managers must be aware of the team process that occurs within a virtual team, they must be able to handle conflict in the virtual environment, they must be able to build trust within their team, and they must apply appropriate leadership strategies. In addition, the virtual program managers must ensure that an appropriate level of technology and resources are available.

Experts agreed that one team-building model was applicable to the virtual environment. As several experts found this model to be highly applicable, this model will be reviewed in detail as a winning strategy for the virtual environment. Several experts identified the Tuckman model as a team model that applies to a virtual program (Curlee & Gordon, 2010; Duarte & Snyder, 2006; Joy-Mathews & Gladstone, 2000; Lipnack & Stamps, 2000). A program manager can apply this model to his or her virtual team in order to predict what will happen during the program and then prepare for conflict during those points in the team process.

The second strategy was how to resolve conflict once it arose within a team. Since the team process might cause conflict within a team, a successful program manager must be able to resolve conflict when it arises. Negative conflict results in both wasted time and lost productivity for a program, because conflict costs the organization money and is a contributing cause of program delays. It is estimated that the cost of conflict, when including ineffective managing of interpersonal situations, conflict avoidance, and lost program days accounted for $20,000 per employee, per year. In addition, 20–25% of a manager's time was spent dealing with team disagreements (Johnson & Johnson, 2000, p. 337). The virtual program manager must identify potential periods of conflict as well as understand strategies to cope with destructive conflict.

The third strategy was how to create trust in virtual organizations. Building and maintaining trust was found to be a significant factor. Many experts identified trust as a significant factor in any successful virtual group. Although trust has already been identified as important, it is critical to recognize this as a successful strategy. Building trust within a virtual organization is a means for success. A trusting organization will always outperform one where there is mistrust and people only looking out for their own goals, rather than the goals of a program.

The fourth strategy was the identification of expertise in programs and prior successful leadership strategies. Successful program managers were those who had expertise in virtual programs, those who were able to mobilize internal support and resources for a program, and the leaders who set high expectations for team members (Curlee & Gordon, 2010; Duarte & Snyder, 2006).

Utilize the Virtual Communication Self-Assessment tool presented in Figure 8.3 to improve your communication within a virtual program.

Name: Date:

VIRTUAL COMMUNICATION SELF-ASSESSMENT

Instructions	Expect to spend about 15-30 minutes with this self assessment. Try not to over analyze the question and answer with what first comes to mind. There are no right or wrong answers.

STEP 1

Do you consider yourself a good virtual communicator?
If yes, list some examples and areas to improve. If no, reflect upon what is holding you back from doing better.

STEP 2

Reflect upon how you can become a better virtual communicator. List areas that you can take action with now.

STEP 3

Consider how you can help change others to become better virtual communicators. In particular, consider how your changes will impact others to change

STEP 4

Set a timeline for these changes in yourself and make a timeline for others to improve as well.
Consider posting the plan and getting others to see it as well.

STEP 5

Design a future state of how you want others to see you as a high effective virtual communicator
Offer to show this plans to others to help get them to help.

FIGURE 8.3
Virtual communication self-assessment.

PHOTOGRAPH 8.2

Finding the forest from the trees. A virtual organization is often hidden under technology, different agendas, and motivations.

9

Technology-Based Communication, Complexity, and Program Management

PHOTOGRAPH 9.0

Flowers have several different complex social meanings. They can be used to celebrate new life, to convey feelings of love, or to mourn the passing of a loved one.

TECHNOLOGY-BASED COMMUNICATION

Social Networks

Technology-based communication relies upon the complexity-based six degrees of separation. This theory is the foundation of the success of social media websites such as Facebook®, Plaxo®, LinkedIn®, and others that individuals and companies may use for personal or professional reasons. The expansions of these networks have caused many companies to restrict these sites; however, some organizations are beginning to understand the advantages they offer in certain areas. These websites create a unique opportunity to help the forward progress of a program.

Complexity theory is based upon the management belief that total order does not allow for enough flexibility to address every possible human interaction or situation. The problem is people are inherently skeptical of less order and flexibility because they believe there is less control. There is evidence to show that virtual programs do better when a program manager uses aspects of complexity theory to lead the program. Six degrees of separation is the belief that people are connected by no more than six degrees, and so if one were to reach out randomly into one's network, one should be able to reach anyone after about six connections.

Recently, social media has touched many lives in our professional or personal network. Many have used instant messenger to communicate to friends, colleagues, or loved ones. Most would agree that it is a fast and efficient means to resolve problems or find the right individual to resolve the problem at hand. In fact, many companies have implemented their own instant messenger system to ensure intellectual property is not compromised. Companies are more apt to create private and public Facebook pages. The private Facebook pages may be for employees to collaborate, while the public Facebook is normally for commercial or marketing purposes.

By leveraging social media sites, a program manager can leverage one's meta-network in a manner to expand one's support and to improve communications. A well-run virtual program verges on controlled chaos since there are many lines of communication and a lack of visual cues. A social network is the same, as one can post or communicate to everyone in the meta-network by leveraging one's normal network.

It is common in business today for a program manager to never have met the team leaders or the majority of the program members. Despite this lack of connection with everyone involved, the program manager is expected to ensure a program meets its expected benefits and strategic goals. The lines of communication ($n(2n/2 + n - 1)$) can become daunting especially since there is the added complexity of no body language which may account for 80% of the message (Roebuck, 2001). Understanding that a virtual project may result in more chaos, it is wise for a program manager to have certain contingencies in order to maintain the balance of control.

Leveraging Social Networking Sites

One practical way to maintain this balance of control is to leverage these social networking sites in a way that supports the program and its components. A program manager should consider creating a social networking site that distributes information. In the past, organizations looked to distribute this kind of information through an intranet or via the corporate website. However, these sites tend to restrict information in a manner that is not productive. Studies have shown that if people need to sign on to another site to garner information, they are less likely to use those sites (Cooke-Davies, Cicmil, Crawford, & Richardson, 2007).

A parallel to this is email spam. Although there are laws against aggressive spammers, there is no doubt that it can be effective as companies continue to use this kind of guerrilla marketing to sell their wares. Just as a good program manager should be able to reach out to the entire team, a good social network will merge elements of one-way communication (examples include the intranet, spam, and the corporate website) with more robust two-way communication (instant messaging, chat rooms). This merging of communication media while leveraging a meta-network will be the vehicle of the future virtual organization.

Trust and Accountability

Trust and accountability are key to allowing monitored complexity to go forward within an organization. Consider when there is a disruption to one of the teams of the program. Given the size and scope of a program, the program manager might not be aware of every situation in every program component. Hence, the program manager cannot automatically insert himself or herself to resolve it unless made aware of the situation. The better decision is to let the team/component resolve it with its network. The team will be more adept at resolving conflict if allowed to resolve it on its own. In fact, depending on when it happens, it may be resolved before the program manager can even find out that there is a problem. If the organization is trained to have the program manager correct the problem, the team will learn to wait rather than to take corrective action. This is not the most efficient type of organization as it forces the organization to have intermittent pauses when the program needs to be continually moving toward the ultimate goal.

Social Teams and Complexity

Entrepreneurial teams that will be able to embrace the growing strength of social networking will be the admired companies of the future. The social networking boom has been supported by the development of many different websites that offer to help people connect to others of similar background or interests. These social teams will be able to reach out to others in order to embrace a greater web of people who could become important stakeholders of the team. These networked people can assist in the development, planning, launch, or even sale of whatever product or project is underway. Consider the strength of having at hand the resources of a host of different individuals with varying backgrounds, all of which have an interest in a program. These people can serve as anything from cheerleaders to beta testers.

Entrepreneurial teams will be able to create teams of fans, followers, and friends who could offer perspective to any program. Imagine the strength that a small team would have if each team member had a secret cabinet of wise advisors. Every aspect of the project and of the team can be improved by invisible advisors that appear to be available to the team.

Entrepreneurial teams will also be able to flex the local community to assist with the goals of the team. A social and gregarious team can leverage everyone around them to be of service to the team. It does not require a great commitment by anyone, but the fact that they can reach out across boundaries to garner assistance from others will certainly give them an edge over a team that is tied to whatever available resources are available. More makes for positive improvements as even Napoleon recognized that victory would favor the heaviest battalions. Entrepreneurial teams will always have heavier battalions when compared to similar-sized rules-based teams.

Rules-based teams will be unable to leverage either a social network or the local community as they will be tied by rules to keep all information, whether important or not, a secret. Furthermore, rules-based teams will never ask for more help for fear of being replaced or supplanted by others. A rules-based organization discourages individuals from asking or giving help to others because by doing so would be to proclaim that their team is either too inadequate or too lazy to achieve their appointed project. Rules-based teams will march to their certain doom rather than ask for help from others.

Entrepreneurial teams are considered open teams, while rules-based teams are considered closed teams. The difference is simple: entrepreneurial teams are willing to open up to others, while rules-based teams remain closed to others. This disclosure to others can help gain support for an entrepreneurial-based team. After all, which team would an outsider trust more—the team that is willing to disclose information about the program or the team that is purposefully secret? The open team certainly has the advantage at communication and trust building not only to the outside world, but to the team itself. If a team has a culture of openness, the entire team will benefit as everyone will understand that openness is the norm and secret behavior is discouraged (or outright punished).

Open teams will also be able to manage the team as replacement members become easier to develop and implement. Consider if a member of a closed team were to suddenly leave the team (a new job). That team member takes all the information of the program or component with himself or herself and then it becomes the challenge of the remaining team members to replace the missing team member. The problem is that the experience of the team member who left is not sufficient by the documentation on the program.

With open teams, since information is more fluid and available, the replacement of a team member is easier. It still might not be seamless, but it certainly will be easier as more information is available and others will be able to fill in any missing information. This becomes a significant advantage as it allows for the efficient replacement of team members (even the team leader) with fewer problems.

When compared to other teams, entrepreneurial teams have significant advantages over rules-based teams. Perhaps in the time before the information revolution, where information was power, a disciplined rules-based team had an advantage, but now where communication is power, a flexible entrepreneurial open-network-based team rules supreme.

The opportunity here is that program managers need to learn to harness these networks in order to become successful in the long term. Too often the program manager clings to what he or she knows and what he or she must know is intervention rather than allowing for a good network to generate good results. A futurist once described that the manufacturing plant of the future would consist of the factory, one man, and one dog. The man's duty is to feed the dog, and the dog's duty is to keep the man from tinkering with the machine. It would appear that the future of program management is more about the program manager letting go and letting a network develop to support the program rather than continually intervening in the belief that the program manager is the sole source to make things happen. In conclusion, six degrees of separation is not just a good idea, it is a critical concept that will drive program management in the future. The sooner that a program manager can integrate technology and social networking into his or her programs, the more successful the manager will be.

PHOTOGRAPH 9.1
Just like a windmill works in harmony with the environment, communication and complexity can work together to achieve impressive results.

COMMUNICATION AND COMPLEXITY

Organizational communication is complex and important to every program. The virtual organization is not about the single steward whose actions control the organization's destiny. Leading a virtual organization is about a single individual who harnesses a personal and business network in order to achieve impressive results. Virtual teams require that leaders shift from a focus upon personal ability to a focus on group results (Cascio, 2000). Individuals who are interested in leveraging the strengths associated with virtual work must increasingly follow this pattern of success through networking in the future.

Given the amount of emails, phone calls, meetings, and other communication happening in any program, a program manager must focus his or her efforts on real communication. Real communication is the only way to challenge this glut of information. Leaders must perform actions that support communication, while including the appropriate tone along with the message of the words. Effective distribution and comprehension of data to all levels of an organization are the essential elements of real communication. Thus, the contemporary organization must learn to communicate better so that messages are received and comprehended. The communication distribution system has emerged as an important force in effective internal and external organizational communications.

Trust in Complex Communication

For a virtual organization to be successful, trust must be part of the foundation of the organization. Trust must include the following elements: competence, integrity, and a concern for everyone involved as well as the environment. These elements of trust are necessary to make an organization successful. Trust must be visible at all levels of an organization in order to achieve greatness. No virtual organization can be successful, let alone effective, without the individuals involved having a high level of trust between one another.

Trust must be addressed daily through competence, integrity, and the concern for program members and the environment. Trust can only form when people have the opportunity to interact enough to the point that people feel that they know and respect each other. Only with respect can come trust, so, one must remember to be reinforcing respect in order to support the atmosphere of trust. Research has concluded that trust is an integral part of a successful virtual team. The challenge for the program manager is how to build trust and confidence with groups of people who lack any form of direct communication. Furthermore, it is hard to build trust within a program when there are difficult times where individuals will have to trust the leader and continue along a path, even when the path is blocked and the goals are unclear.

The program manager must learn to maximize the interactions so that each one counts toward building trust. One needs to consider if each interaction (think of an interaction as an email, a phone call, a program-wide memo) is building or destroying trust. The interactions that create trust are desired over those that do not create trust. One other important consideration is the actions of the program manager. The program manager must remember that the actions taken will be remembered and retold to others who were not present. If the program manager takes a strong stand to protect the environment and to make that part of the goals of the program, then people will remember that; if the actions of the program manager support the statements, then people will remember that and tell others about those interactions.

Beyond what can be done to build trust through the direct actions of the program manager, the program manager can also build trust and confidence through team (program) success. Success leads to more trust and communication as people have something positive to celebrate and to discuss. A successful program suddenly has new fans and generates trust in the program. Fans of *The Deadliest Catch* will recognize that nothing brings a crew together like catching full pots of crabs. All differences and issues fade away as the group works to haul up crabs and reap a big paycheck.

The program manager must recognize that if program success is the panacea for trust, then program failure is its nemesis. Program failure will undermine efforts of building trust and communication faster than any other factor. When a program is faltering, people will try to distance themselves from it in order to avoid being associated with the stink of failure. This does not always happen, but it happens more often than not.

The reason for this is that real trust is generated when a program manager completes and honors his or her promises and commitments. Achieving milestones is an organizational promise, and the currency that rises or falls is the trust in the program manager. The program manager must understand that once a positive or negative impression is formed, it is hard to change. Since perception rapidly becomes reality, the program manager needs to actively promote and celebrate publically as many successes as possible. Consider looking at one's ability as a face-to-face manager by using the self-assessment tool presented in Figure 9.1, or by using the Virtual Program Review in Figure 9.2.

Benefits and Challenges of the Virtual Organization

A virtual organization is a group of individuals who work across space, time, and organizational boundaries that are brought together to perform interdependent tasks, united by communications technology and a common purpose. Furthermore, a virtual organization is usually a geographically distributed society bound by a common goal and whose members communicate via information technology.

Name: Date:

FACE TO FACE COMMUNICATION SELF-ASSESSMENT

Instructions	Expect to spend about 15-30 minutes with this self assessment. Try not to over analyze the question and answer with what first comes to mind. There are no right or wrong answers.

STEP 1
Do you consider yourself a good face to face communicator? If yes, list some examples and areas to improve. If no, reflect upon what is holding you back from doing better.

▼

STEP 2
Reflect upon how you handle conflict. Consider how you can better address conflict with face-to-face communication.

▼

STEP 3
Consider how you can become a better face to face communicator. In particular, consider how your changes will impact others to change

▼

STEP 4
Set a timeline for these changes in yourself and make a timeline for others to improve as well. Consider posting the plan and getting others to see it as well.

▼

STEP 5
Design a future state of how you want others to see you as a high effective face to face communicator Offer to show this plans to others to help get them to help.

FIGURE 9.1

Face-to-face communication self-assessment.

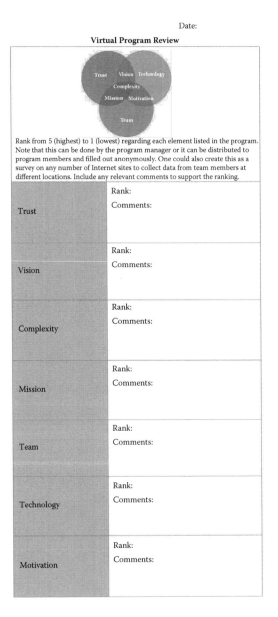

FIGURE 9.2
Virtual program review assessment.

The virtual organization can be a vehicle for programs by offering alternatives and additional business and environmental justifications. Business justifications include documented increased productivity of virtual organizations, better access to global markets by crossing national boundaries, fewer managers, and the positive environmental impact of reduced automobile emissions (Cascio, 2000; Elkins, 2000). Business shortcomings to virtual programs include the high cost to set up and maintain individual employees in home offices, communication challenges, as well as workers feeling isolated from the organization; however, these shortcomings are more than offset by productivity gains.

Increased productivity has been measured by internal IBM studies with regard to virtual employees. Productivity increased between 15 and 40% for virtual employees when comparing them to typical office employees (Cascio, 2000). Productivity increases in customer service have also been documented. Arthur Andersen cites an increase in 25% of face-to-face time with customers in those salespeople working virtually (Cascio, 2000). Other studies have documented problem-solving productivity gains in individuals working virtually.

Virtual organizations require fewer leads and managers. Virtual organizations make use of shared leadership, where every individual is required to perform some role of leadership. Leadership will shift according to the requirements or objectives of the organization, as team members each have to carry the burden of leadership occasionally.

Reduced automobile emissions are a direct environmental impact of telecommuters. If more companies were to adopt virtual teams, vehicle emissions would drop dramatically. In an example from Georgia Power, 150 telecommuters were able to reduce automobile emissions by 35,000 pounds annually (Cascio, 2000). If this example is scaled upward, the environmental impact is dramatic. If there were 100 million U.S. workers telecommuting, the reduction in emissions would be over 23 million pounds of emissions across the United States. Virtuality will redesign society with fewer skyscrapers and fewer commuters, making for a quieter world (Handy, 1995). The opportunity to reduce emissions on such a scale can have a significant environmental impact.

Communication challenges can arise as people lack the robust personal contact of day-to-day physical proximity and interaction. Technical challenges can occur that can serve to exacerbate communication issues (Cascio, 2000; Karl, 1999). Working across great distances can also involve working with people in different time zones, countries, cultures, and languages (Karl, 1999). All of these issues will create communication challenges that do not occur in a traditional office environment.

Isolation issues can occur in virtual teams because these teams lack social interaction, which typically occurs around the water cooler in traditional companies. These social interactions include, on the positive side, the passing on of vital organization information of the activities of others, and on the negative side, the passing on of organizational rumors. Both types of communication are necessary as even the seemingly negative interactions create an atmosphere of connectivity to the organization. In order to avoid such instances of isolation, team leaders must harness technology to amplify the necessary community contact to keep individuals in the loop (Cascio, 2000; O'Connor, 2000). Leaders must communicate and create relationships based on trust, because leaders who do not remain in contact with their groups have the potential to lose the trust and effectiveness of their teams (Cascio, 2000; Handy, 1995; O'Connor, 2000; Platt, 1999).

SUMMARY

Program managers are being pinched from all sides for results, and project managers gravitate toward using the same individuals and resources for similar tasks; however, over time even these kinds of relationships can fail. A finely tuned machine will ultimately fail if maintenance is not regularly done. This is even more apparent in people when relationships are ignored or taken for granted for too long. Program managers might not grasp the importance of trust on a program when everyone is focusing upon results without looking further to find out why the people were not successful. Budgets, deadlines, stakeholders, and other situations become the daily priority, and then the program manager finds himself or herself without time to devote to building trust. Trust is built over time and is done through positive communication. There is no sure way to create trust, but if a virtual program manager is attentive to the communication and social needs of the project team and social needs of the program team trust is established throughout the program. If a program manager has time for making calls to find out status, the program manager has time to maintain a positive and trusting relationship with everyone that he or she contacts. People agree that trust is earned, so the virtual program manager must always consider what he or she does on a daily basis to continually earn the trust of others.

PHOTOGRAPH 9.2

Rainbow over building. When the conditions are right, the rainbow will appear. Just like when the conditions are right, a successful program will emerge.

10

Course Materials #3

PHOTOGRAPH 10.0
MQ-9 Reaper. (Courtesy of the Air Force News Service.)

CLASSROOM MATERIALS

Case Study #3—Communication in Program Management

Military Technologies Inc. vs. Guidance Systems LLC—Part One

Jack Smith works for Guidance Systems LLC. He enjoyed his position as contract manager for the Military Technologies Inc. account for 10 years. Guidance Systems LLC is currently the subcontractor to Military Technologies Inc. for guidance systems for the new generation of military support ground vehicles. These vehicles are state of the art and have considerable technology on board to move the vehicle as well as to notify the driver of friends and foes. This vehicle is under development for ultimate use by the U.S. military. Jack understands that there has been a lot of work put into this agreement, and he is very proud of how far it has come along.

Jack had worked the contract for three years when it came time for a contract modification to start to integrate some proprietary technology that Guidance Systems LLC wants to include in the next phase of this program. As he prepared for the contract meeting, he sat down with the new sales manager from Guidance Systems LLC, Graham Garnet.

"Graham, I know you've never dealt with Military Technologies Inc. before, and I want to make sure that you understand what you're up against. They are very demanding and they never want to hear no," Jack explained.

Jack leaned forward, "You don't know Military Technologies Inc. They want everything yesterday, and if you show any weakness they are going to hand you your lunch and ask you to pay for it. This proprietary technology appears to be the solution to a lot of problems that we have been having, so it is very important that we get this deal done." Jack handed a list of other items that would be great to include in this new contract revision. The largest one would be for an exclusive for this technology to give the company a leg up on the competition.

"You think I'm going to give away the company?" Graham asked. "I have dealt with a lot tougher companies than this little subcontractor. I would think that you would know that I have more experience than to just hand them the store." "It isn't that at all, I am sure that you have worked with other subcontractors before, but we really need to get an exclusive on this proprietary technology," Jack responded. "Over the past few years, we've learned that Military Technologies Inc. underestimated the costs of this project, and they have been strapped to make it work. They are dedicated to the program, but in the short term they are really pinched to make a dime. We try to work with them on some change orders, but we need to make sure that the program is a success and on budget. This program means a lot to the company, and I believe that this is going to keep the company afloat for years into the future."

Graham responded, "I brought along some interesting data on other similar technology that is available through others." He passed the file across the desk to Jack. Jack looked briefly at it and shook his head no. "None of these options are going to work and they know it." Graham disagreed. "If Military Technologies Inc. shops around we can find something else, and if Guidance Systems does not wake up to this fact, they are going to find themselves replaced."

"Graham, none of those options will fill our needs, and we don't even have an agreement with any of those other companies! It will take time to build up a relationship, and it would take them time to get to the same level as Guidance Systems and that time is going to cost money." Jack was visibly annoyed and found that Graham was ignoring him and his experience with this account.

"Look, Jack," said Graham, "I know Military Technologies Inc. well enough to know that if we start talking to others that they will go sell their technology to someone else. They are a small company but I know that they still need to make money. If they know that this new system is better than the rest, they will be working to sell it to others," Graham said. "When we go to negotiate with this company, we really need to have a strong position to get the best price. I think that you want to show up and beg for a good deal. I do not agree with your approach because they have more to lose than we do."

"I think that this discussion is over," said Jack. Jack then stood and walked toward the door. "Honestly, Graham, I think that you have more of the interests of Guidance Systems LLC, rather than Military Technologies. You need to keep your focus on the company, because we are here to make money. I cannot help it if they did not plan properly and they are struggling. If they are struggling as much as you say, they will be happy to get more business from us."

Later that day, each of these people comes to you as the program manager to complain about each other. They tell you more or less the same story about how it went down. You know that you have already promised the client that your company has a great relationship with Guidance Systems and that you are just about to lock this proprietary technology down and it will be the cornerstone of the next generation of this project. They do not know that you have promised this to the client, but now you feel that their actions are really going to make this promise difficult to keep.

Case Study Questions

1. How should Jack, as the contract manager, and Graham, as the sales manager, prepare for the negotiating session with Military Technologies Inc.? Explain how you will direct them both to ensure that they achieve the necessary goals.

2. What concerns do you have and how will you address them in the negotiation? You know that both of them must be present, because you have already told the client that there is a new sales manager who is the new go-to person. You will not be present because you will be at a supplier meeting with the U.S. military. How will you ensure that they are able to close the deal?

3. As a leader, what will you do to remedy this situation quickly and to get them both on track for this upcoming negotiation? Describe your communication plan for the two of them.

4. Explain how you will improve communication in the program to make sure that these two continue to work together in the future?

PHOTOGRAPH 10.1
Finding the right answer among the forest of sameness.

SECTION QUIZ

Section 1

Multiple Choice

1. Which styles of leadership theory are more appropriate for a complex situation?
   a. Complexity leadership theory; transformational leadership theory
   b. Complexity leadership theory; autocratic leadership theory
   c. Complexity leadership theory; Machiavellian leadership theory
   d. Complexity leadership theory; flexibility leadership theory

2. Social media is based on what complexity theory?
   a. Anthill theory
   b. Butterfly wing theory
   c. Six degrees of separation theory
   d. Chaos theory

3. Complexity theory states that humans are:
   a. Closed systems
   b. Chaotic systems that cannot be relied upon
   c. Systems of opportunity
   d. Open systems that are affected by their surroundings

Section 2

True/False

1. Facebook, Plaxo, and other social media sites are examples of the butterfly theory.
   a. True
   b. False

2. Transformational leadership is a type of leadership that should be used by program managers.
   a. True
   b. False

3. Creativity occurs at all stages of the program.
   a. True
   b. False

4. A program happens in a linear fashion.
   a. True
   b. False

5. Program managers demonstrate leadership by communicating to the program's stakeholders.
   a. True
   b. False

Section 3

Answer Key

Section 1

  1. A
  2. C
  3. D

Section 2

  1. B
  2. A
  3. B
  4. B
  5. A

DISCUSSION QUESTIONS

Short Discussion Questions

Discuss the similarities between the Transformational Leadership and Complexity Leadership theories.

Discuss the various Modern Leadership theories and how they would or would not apply to complexity.

SUMMARY

Program managers not only must learn to leverage the traditional organization, but they must also be savvy with regard to using social networks and other technologies that are available. It is important that the program manager stay in touch with the program and with all the stakeholders involved. A program manager must be able to multi-task in a manner to keep the program, the program team, and the customer all in focus. It is not easy to achieve, but by leveraging different communication technologies, the program manager has the possibility of success.

PHOTOGRAPH 10.2
The road less traveled. Program managers need to learn how to take the road less traveled in order to achieve unique success.

11

Complexity-Based Program Management

PHOTOGRAPH 11.0

Applying program management and complexity may appear to be as difficult as climbing up a glass staircase; however, with knowledge, experience, and practice it can become seamless and successful.

APPLICATION OF PROGRAM MANAGEMENT AND COMPLEXITY

Experienced and successful program managers intuitively jump from linear to non-linear (complexity theory) thinking at the appropriate points in the program. Scheduling and cost management normally will be linear activities, while conflict management would be in the realm of *edge of chaos* or *non-linear* thinking. New program managers may attempt to handle all situations in a traditional mentality of step two only comes after step one, and situations can only be handled in a strict chain of command. What these inexperienced or rigid program managers may find is that the program personnel become disenchanted because there is no perceived trust to handle situations at lower levels. Or the program manager may find that he or she is doing nothing else but the program, which may mean handling crisis upon crisis. Complexity is not a simple delegation. Complexity is about empowering teams and individuals to act.

Programs normally last for many years since they have a strategic implication. Programs also have component projects. Some may argue the three highlighted programs throughout the book are not programs, but no one would argue their complexity. Let's review each for its worthiness as a program.

The attack on the United States on September 11, 2001, was coordinated by several hijackers (the exact number is unknown, at least in unclassified articles) crashing commercial passenger jets into large commercial buildings, one federal building, and one unsuccessful hijacking that crashed in a field in Shanksville, Pennsylvania. Once the U.S. leadership realized the country was being attacked, the governance structure was quickly put into place. An overall authority was assigned by the President of the United States. The strategic goals were to defend the country and prevent any further attacks. The component projects were to land all planes and divert any that were not in the continental United States, put the Air Force and Navy Air Assets in the air and on alert, have the Coast Guard protect the borders, and find the hijackers. There are strategic goals and component projects. Some projects lasted longer than others, and the strategic goals were refined as time passed.

The Gulf oil disaster was initially viewed by BP as a project. The steps taken by BP were tactical in nature—there did not seem to be a strategic mindset except to minimize the bad press and the monetary damages against the

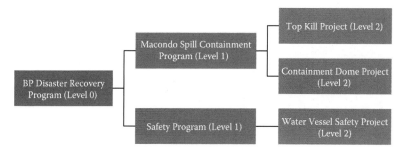

FIGURE 11.1
Hypothetical WBS.

company. As time passed, BP implemented a governance structure, most likely demanded by the U.S. Coast Guard, who took control of the day-to-day activities at the height of the spill. The Coast Guard's program manager and de facto program manager for BP until BP could show stability was Admiral Allen. In essence, anything that BP did that was related to the disaster was approved by Admiral Allen or his governance staff.

While there were no published organizational charts or work breakdown structures (WBSs) of the overall program, a program of this magnitude would have component projects, component programs, and component activities. Figure 11.1 presents a high-level hypothetical WBS for a portion of the BP oil disaster.

Japan was struck by two natural disasters almost simultaneously which devastated many towns, killed many people, destroyed families, devastated 20 of Japan's prefectures, and caused a nuclear power plant to have several meltdowns. The Japanese government has emergency services in place to respond to quakes and tsunamis but not to the devastation presented by the 9.0 quake and tsunamis that were as high as 133 feet. Entire towns disappeared, infrastructure disappeared, and logistics was shattered as the northern part of Japan has much of the manufacturing for the country.

Japan at that moment could not even think about the country's long-term stability. The government had to quickly respond to the needs of the survivors, look for survivors, bury the corpses (which broke culture), deal with the mounting urgency of the Daiichi nuclear plant, coordinate the domestic disaster relief equipment and the international equipment, coordinate personnel domestically and volunteers from other countries, take care of the newly orphaned, provide shelter, and the list continues.

All these projects and programs have to do with the survival and reconstruction of a nation, which is the strategic goal. Bringing them together under one program provides synergies and allows the program manager(s) to establish benefits that the individual projects/programs would not be able to realize individually.

These three programs demonstrate how complexity can be used to help a program achieve momentum via non-linear thinking.

- U.S. FAA Towers and September 11, 2001—Individuals overcoming insurmountable communication and coordination complexity
- BP Gulf oil disaster—Complexity with cleanup efforts, coordinating vessels, maintaining safety, and capping the ruptured line
- Japan's earthquake/tsunami/Fukushima Daiichi nuclear disaster—Complexity with mass casualty, nuclear disaster, foreign government assistance, and still running a government

Once the decision was made not to use command and control to land all planes in the United States on 9/11, the personnel in each control tower had to determine how to land the planes. Remember each control tower's situation was different, and the environment was dynamic. Planes were entering and exiting the tower's control space, and there may not have been enough room at the airport to land the planes flying in the control tower's air space. The control tower's personnel had to negotiate to land these planes at other airports, understanding that these planes may still have hijackers on board. At the time, the United States did not yet fully understand the situation.

The control towers also had to deal with Canada and Mexico to divert some aircraft that may not have space to land in the United States. Some flights that were heading to the United States but could not be turned back to their destination such as flights from South America, Europe, and Asia may have all been diverted to Canada, Mexico, or Iceland, returned to their point of origin, or returned somewhere in between their destination and point of origin. Remember this was done dynamically. Also, for planes that did not conform to the instructions of the air traffic controllers, military jets had to be scrambled. Although the public may not ever know (due to need-to-know and classified information), it is hard to fathom how many more commercial aircraft may have had hijackers, how many more military aircraft scrambled on civilian aircraft, and how many safety issues there may have been because of the sheer chaos between and within airports.

To put it in perspective, on any one day of aviation, there are approximately 87,000 flights in the United States. These include commercial passenger flights, air cargo, private planes, general aviation, and military. In relative terms, it would take 460 monitors to show all the aircraft being monitored at any one time during the day (National Air Traffic Controllers Association, 2012).

There have been simulations done on the 9/11 terrorist attacks and the U.S. response. To date, no other simulation has yielded a better response than the one undertaken, which utilized complexity. Many of the other scenarios demanded command and control or a hybrid of command and control. There was no freedom to work the situation as needed with the most up-to-date information with those closest to the event—those in the control towers directing the planes.

BP staff and management addressed the Gulf oil spill in the traditional manner outlined by the company's published spill response plan. Once the crisis occurred, BP moved forward with an organized response as outlined by the plan. The plan was underdeveloped to address a situation of this magnitude, and the situation quickly overtook the individuals involved. The process showed that initially the response seemed to follow a normal, formal risk plan, but it was improbable that such a system would be successful. Once the magnitude of the spill was accepted, BP and other organizations needed to mobilize a vast armada of equipment, people, and materials to combat the spill. The spill was larger and more complex than anything that had been seen prior in the United States, and so it became necessary to develop new systems and processes in order to be successful in the shortened amount of time.

Four critical findings were discovered regarding the BP response to the Gulf oil spill:

1. The amount of people and equipment involved in the process was an unprecedented buildup in an amazingly short period of time, a situation that was not covered in the existing BP response plan. This required a complexity-based system to mobilize and de-mobilize the vast resources necessary to combat the oil spill.
2. Traditional risk planning was abandoned in favor of on-the-spot response where management was able to respond dynamically, as the BP spill response plan was very poorly conceived and written and unable to cope with the magnitude of the Gulf spill.
3. BP now acknowledges the need to move the center of operations from a static base in Houston (as outlined in their response plan) to on-the-spot management empowered to make rapid decisions as new information came to light.
4. Communications locally were identified as critical to the success of the spill cleanup (Curlee & Gordon, 2010).

There is sufficient data to support the buildup of people, material, and equipment in a very short time. Part of this buildup has been from the poor initial estimates of the magnitude of the spill. Early estimates were showing that the maximum amount of the spill was 5,000 barrels (National Commission, 2011) which was shown to be a completely inaccurate forecast. This initial forecast hampered the increase of the response, because the BP spill response plan requires that an estimate be used as a first step (BP, 2010). These rigid elements hindered the initial attempts to contain the spill. The estimates were entrenched as fundamentally accurate which caused further delays in the deployment of additional equipment and material.

Since the original high estimate was 5,000 barrels per day (which was externally confirmed outside of BP), BP responded and deployed dispersant material to handle the 5,000-barrel-a-day spill. When the initial deployment of dispersant was found to be inadequate, the only two possibilities were that the dispersant was defective or there was substantially more oil in the water as to make the quantity of dispersant ineffective. Because it was assumed that the estimate was correct, precious time was wasted because the dispersant was being blamed as being defective. Once the dispersant was found to be in good working order, the only possible alternative was that the 5,000-barrel estimate was inaccurate.

At this juncture, BP should have considered both possible options instead of pinning hope on the dispersant defect. Also the two options were the dispersant was bad or the estimate of the number of barrels was incorrect. BP stood behind the number of barrels being correct and the dispersant. The decision to look linearly at the problem rather than work on multiple possibilities shows how the limited risk plan hampered the initial attempts at spill containment. Early failures like this helped BP move from following the entrenched risk plan to a more dynamic method of addressing the issue. The sad point is that it took so long for BP to realize this deficiency, and only later did they accept the magnitude of the spill and moved to respond appropriately.

"Complex systems almost always fail in complex ways" (National Commission, 2011, p. vii). In the end the risk plan developed by BP was fundamentally flawed (if one were to ignore the obvious glaring mistakes, such as information on how to handle sea lions and walruses in the spill plan, and the fact that the environmental consultant identified was dead years before the spill plan was submitted) in that the plan spends too much time discussing a very linear approach to an oil spill (National Commission, 2011).

The plan includes pages of flowcharts on how and when to deploy dispersants for an oil spill. Furthermore, the deployment of dispersant is only considered if the oil is heading toward shore or colonies of sea birds. No other wildlife is considered as part of this process. If the flowchart allows for dispersants to be used, there is an internal requirement where approval must be sought in order to apply dispersants. This linear requirement seems to ignore that on-the-spot action might serve better than a hierarchical and linear response. Only after the spill is flowing out of control and national attention has been attained does BP start to move with more alacrity.

It is interesting to note that even after BP submitted paperwork as the responsible party on April 24, 2010, the mobilized response was quite small. On April 28, 2010, when the U.S. Coast Guard identified BP as the responsible party, the mobilization effort began in earnest. Within two days, BP moved from 500 people deployed (mostly for call centers to handle claims) to 2,000 people and 75 vessels deployed on April 30, 2010. It is clear that a shift occurred within BP management that the Gulf spill was no longer a simple matter of closing the well, applying dispersants, and deploying skimmers. The leak was larger than anyone had ever anticipated, and the process to clean up the spill and to close the well would take months. No simple risk plan flowchart would be able to cover a situation of this magnitude.

Japan's government knew how to deal with tsunamis and earthquakes; however, dealing with two natural disasters of disproportionate magnitudes that were so far reaching taxed the government's response system. Compounding the issue was the Daiichi nuclear disaster. This series of events will be studied for years to come, but what is known is that there were many areas on the edge of chaos and those that had slipped into chaos. Many were struggling to survive, and others were trying to save others from certain death (for instance if the Daiichi reactors had gone in to total meltdown). There were those who were able to overcome the shock of their situation and bravely gather survivors and help to ensure their safety until proper authorities arrived. This could be seen throughout the various devastated prefectures. Companies that could match survivors with relatives helped as soon as possible. The Japanese government appeared to be helping as quickly as possible, but TEMCO, the owner of the Daiichi plant, was not forthcoming. TEMCO initially took the same stance as BP in the oil disaster. The company wanted to protect the reputation of the company, minimize the losses to the company, and provide a campaign that twisted the actual disaster that was happening with the reactors within the plant. Once explosions started happening and radiation was leaking, it was hard to mask the truth.

How do these three programs help the program manager understand complexity for his/her program? For the 9/11 program and the Japan disasters, the individuals involved were taught how to deal with the unexpected. Air traffic controllers have to respond to aircraft with emergencies and planes entering restricted air space. There is not a process or procedure for every foreseeable event that a plane or pilot may encounter. Because of the events of 9/11, many things have changed within the cockpit of the commercial aircraft. Pilots no longer cooperate with a hijacker, the door to the cockpit is locked and the door is fortified, and air marshals are randomly assigned to aircraft. While changes were made, complexity maintains that there is more than one answer to any given event.

For Japan's tragic natural disasters, the communities had been prepared to respond. All knew to go to higher ground as soon as the earthquake hit. Most of those who did survived. Those who did not perished. Many who did not go to higher ground were the elderly who did not have anyone to help them. In many places the first tsunami hit within 10 minutes. This is not much time to respond. The survivors knew to wait and started to help as soon as possible. These individuals had been provided the tools to be leaders in the areas of chaos.

BP leadership took much longer to equip its program team to react to complexity. Why? The BP structure was established for command and control. Not until the U.S. Coast Guard took control did the harnessing of complexity take over. The Coast Guard equips its cutters and units to act independently and forced BP to start doing so as well. When this happened, complexity momentum started to occur.

In business, the program manager needs to equip his/her program staff and component activities to recognize complexity and to have the courage to deal with the *edges of chaos* opportunities. The program manager must lead from the front. Shying away from complexity opportunities will demonstrate to the team that tackling chaos is not encouraged. The program manager needs to assess the success/failure of these opportunities. They need to be learning experiences for the entire team, not times to berate or demoralize. Remember that there will be failures when complex situations are taken on, but let the team fix the failure. Stepping in may actually hinder the team, so tread lightly in these situations.

Complexity offers explanations for social systems without limiting the explanation to a single variable understanding. People want to have an explanation for all matters within a program in order to replicate the solution at a future time. Complexity is not about creating a single style solution. Complexity is about accepting that there are multiple solutions.

PHOTOGRAPH 11.1
Coastal marshland was of high concern after the Gulf oil spill as many species use these areas as breeding grounds. The impact in such areas is still not known. Coastal marshlands are also complexity-based systems where experts understand that they are important areas for breeding; however, it is not understood what could happen to hundreds of species if these marshlands were destroyed.

12

Applied Complexity and Program Management

PHOTOGRAPH 12.0
Integrating technologies and program management may seem as difficult as using solar panels on a cloudy day; however, with time and practice one can successfully leverage technology.

INTEGRATING TECHNOLOGY
INTO PROGRAM MANAGEMENT

Complexity theory in some respects is a stimulus behind social media. Six degrees of separation is a sub-theory of complexity and is the driver of the social media sites. Most of us would like to think we are just six people away from being buddies with a rock star, the CEO of a company, or the person who can get you the next job. Whatever the enticement is, social media has taken off.

Several of these social media sites were created by college students, including Facebook founder Mark Zuckerberg. These students wanted to find a better way to collaborate with their fellow students. Some may have been introverts and did not make friends easily, others may have wanted help with a certain subject, and others may have wanted something else, the list is endless. Imagination and creativity are the only limiting factors on what can be done with these social media sites. The darker side of humanity has also been seen on these sites.

Businesses initially pushed social media sites to the side. Companies do have a legitimate concern that intellectual property may be lost, there may be a concern as to the way employees may conduct themselves, and there may be the concern that sensitive data other than intellectual property may be lost. Some companies did start to see the potential and started to mimic the sites internally. This typically did not maintain traction in many companies.

The reasons were varied. The platform the company may have created probably was not as nimble as those that were created for social media. The majority of the employees most likely were not avid users of the sites when social media first become popular. Those employees who did use the social media sites at home were hesitant to use it for work purposes. Companies found that maintaining these platforms was expensive, and many companies were not quite sure what to do with them. Most companies found that using the actual commercial social media site with security protections or using an unprotected site for commercial purposes was the best of both worlds.

Prior to social media sites, companies adopted instant messenger, wikis, blogs, texts, video teleconferencing, and some now even have introduced Twitter. Some of these technologies come and go depending on how well the company is able to integrate them into the culture of the company. The technology savvy of the company, the median age of the employees, the appetite for change, and the amount of churn will drive how all the various technologies are accepted. The program manager should have an organizational change management work stream in order to help end users with accepting and understanding the program, and timing training to coincide with the implementation.

Three of the most studied natural and manmade disasters have been discussed throughout this book. Technology was integrated into these three programs at various degrees and will be discussed further:

- U.S. FAA Towers and September 11, 2001—No state-of-the-art communication technology was used.
- BP Gulf oil disaster—State-of-the-art communication technology to the commonplace (such as face-to-face communication, walkie talkies, mobile phones, line of sight, and others).
- Japan's earthquake/tsunami/Fukushima Daiichi nuclear disaster–State-of-the-art communication technology to the common place.

The terrorist acts committed on September 11, 2001, led to a program that U.S. leadership had never had to deal with before. Commercial passenger jets were hijacked and being used as suicide vehicles to fly into high-valued targets, such as the New York Twin Towers and the Pentagon. A third plane was headed for another strategic location in Washington, DC but never made it. Leadership was facing unanswered questions. Were there more hijackers? If so, were the hijackers only on the East Coast of the United States? Were any hijackers coming in on international flights? The decision was made to ground all flights and divert international flights.

There were no processes in place for this scenario. There were flights coming from Canada, Mexico, South America, and flights over the Atlantic and Pacific that had to be diverted or turned around and sent back home. Then there were planes flying all over the United States that had to be identified and told to land. Airports had to prepare for an influx of planes that were never supposed to be there. International flights landed at airports that may not have sufficient resources to handle customs and immigrations.

Initially, it was decided to control, monitor, and land all aircraft from one central authority, but the FAA soon realized that air traffic control is a series of spider webs which are handed off at appropriate points with some overlap. Once the emergency was declared through the system, each tower coordinated the landing and diverting of planes that was in their responsibility at that time. Granted this is a somewhat simplistic view of what took place, but once the decision was made, communications on the edge of chaos took over.

Each tower and its personnel communicated with each aircraft (civilian, commercial, military, foreign) and other towers until all aircraft had landed. Tower personnel did scramble fighter jets on occasion, as some aircraft did not respond to the Tower's command. In this heightened state this is understandable. What should be noted is that standard technology was used throughout the entire event. There was no time to bring in an untested method or a technology that might have been able to view every aircraft flying over and into the United States. The people in the middle of the chaos did what they had to do with what they had.

Scenarios were run after the fact, and it was found that the best way to handle a situation of this magnitude in the future is to do it the same way. In other words, to let the "on-scene commander" handle his or her area. That is exactly what each tower did when identifying and grounding each plane. Were there more hijackers on any of the other planes? We will never know; however, it does not matter because of the thinking out of the box of each of the control tower personnel who landed those planes safely in a chaotic environment and in a record time.

The BP oil disaster was not as quick to respond to the complexity, and communication technology was advanced in this case. BP had an oil rig in the Gulf of Mexico that had a catastrophic failure that resulted in an explosion with loss of lives and a massive oil spill. Initially, there was much finger pointing as to who was responsible for the disaster. With major operations of this sort, there are many companies involved, and BP wanted to shift the blame. BP quickly settled with the families of those who died, and again BP viewed this in a manner to minimize the company's loss. This was a marketing and political disaster for BP and meanwhile thousands of gallons of oil were pouring into the Gulf.

Once the United States put the onus on BP, there was a shift in its approach to handling the situation. After the initial disaster, BP was taking a tactical approach of putting out fires (figuratively), circling to protect the reputation of the company, and minimizing financial damage to the company. As the company realized the disaster was not subsiding but was indeed becoming worse, a program-focused approach was taken.

The program took some missteps, but this can be expected since the governance had to be put together so quickly and there were so many stakeholders with such diverse needs. As the program began to take shape, it became clear that command and control could not be centralized because of the diverse set of component projects. The need for 24/7 updates and safety was compelling because of the number of vessels (airborne and waterborne). The waterborne vessels ranged from small craft to large tankers and many times they were in close proximity, within feet of each other. This could be compounded with a helicopter causing the small craft to shift because of the tremendous force given off by the helicopter's rotors.

For this program, advanced technology was a must. There were over 5,000 vessels at the height of the BP cleanup operations. Many of the vessels were equipped with Automatic Identification Software (AIS). This allowed the "on-scene commander" to understand where his/her vessels were in relation to the other vessels in the area. Additionally, it relayed information to the overall commander for all vessels and to the onsite safety officer. Everyone was also expected to use the first line of safety, his or her eyes, scanning constantly to make sure that each person and vessel was safe, as technology can always fail.

Traditional technologies were used on the program as well. There were dashboards, the U.S. Coast Guard used its technologies to communicate, hand radios were used for the many volunteers, email, the U.S. Navy used its communication, and commercial vessels had their communications. The program also established a mechanism to communicate with the public. There were radio and television advertisements. There were also ways to handle the media.

During all of this, communication technology was being used outside of the program which affected the program. There was a social media site dedicated to scientists where a problem would be posed and various scientists would debate one or more conclusions. During this trying ordeal the mediator of this site offered scientists the problem of the BP oil "gusher" and its solution. These scientists were from all over the world and are some of the most brilliant minds living today. When the solutions were offered to the U.S. President and to BP, they were all discounted or ignored by both parties. Could it be there was an overwhelming amount of communication, or that scientists do not command the same attention as the actors and politicians who became involved? When a communications plan is developed by the program manager, the plan must have response strategies for the media, dignitaries, internal and external stakeholders, and other communications that might affect the program.

Japan has a history of earthquakes and tsunamis. Earthquakes have been known to trigger tsunamis. In 2011, Japan had a massive earthquake which triggered tsunamis as high as 130 feet. These events happened almost simultaneously and wiped out the infrastructure of the areas affected and killed tens of thousands of people. When it happened it created a chain of reactions at the Daiichi nuclear power plant that caused three of the reactors to partially meltdown. Had the plant been affected by just one or the other event, Daiichi most likely would have survived with some structural damage and possibly some minor radiation leaks.

After some investigation, TEPCO's (Japan's Power Company) leadership acknowledged poor preventative maintenance, maintenance, and management at the power plant. The power plant was not designed to withstand a one/two punch—an earthquake followed by a tsunami. There were many within the power plant who did some heroic deeds at the risk of their own safety to save those in the surrounding area, because they understood the gravity of the situation even before management was willing to say anything.

Japan's government had to establish a difficult program to reestablish a part of its country and stabilize a nuclear power plant. While the Japanese government and TEPCO leadership were stating everything was alright with the nuclear power plant, other countries' governments were telling their citizens to return home from Japan. Japan was slow to react to advise its own citizens to move from the hot zone around the plant, although a U.S. carrier docked in Yokosuka over 100 miles from the Daiichi plant detected radiation 12 hours after the disaster. This is advanced technology in action, but it was ignored.

Technology was provided by France and the United States to help with the nuclear plant. Numerous other countries, including France and the United States, provided technology, people, and aid to help with the initial disaster relief and post-disaster help efforts. Communication was extremely important and complex in this disaster because of the severity, the amount of the population affected, and the impending nuclear situation. The Japanese government needed to establish a communication tactic similar to BP and 9/11 but with a focus on safety and the needs of the people. There were many lines of communications and many edges of chaos. There were many miscues with communication. Infrastructure was gone in this area of Japan. Survivors can be expected to be in shock, and post-traumatic shock may ensue if proper medical care is not established. All communications' infrastructure was overwhelmed, but there may have been other avenues available. As mentioned before, a U.S. carrier was approximately 100 miles from the epicenter of the disaster. A carrier is normally the center of the world when a group of naval ships deploy.

The carrier may have served as an alternate communications and logistics infrastructure for the area affected until the Japanese government was able to cohesively bring all parties together. This or another strategy may have been utilized but to date is not readily available in open source literature.

The lessons learned from these three programs are that technology needs have to be evaluated for the program and component projects. What may work for one program may not work for another program. Advanced technology may be required for a program for safety concerns, while another program would flounder with state-of-the-art technology.

Most are not working under the time constraints, disaster conditions, and tremendous stressors that the above programs placed on their program managers. We normally have the time to plan the program and the technology that makes sense or to adapt the technology as the program matures or changes, as component projects come and go, and as benefits need to be adjusted to meet the strategic goals of the program and ultimately the company.

PROGRAM VERSUS CONSTITUENT COMMUNICATION AND COMPLEXITY THEORY

Senior project managers and program managers have always realized that stakeholder engagement and communication are part of the project management discipline. This was formally confirmed in 2013 when the Project Management Institute (PMI) published the fifth edition of the PMBOK and included a tenth knowledge area, Stakeholder Management.

PMI's core team writing the Program Standard (2013) also acknowledged the importance of stakeholders. This core team realized at the program level it was not managing the stakeholders but rather it was a matter of engaging them. The section in the Program Standard is aptly titled *Program Stakeholder Engagement* (PMI, 2013b). An individual cannot be managed, as the standard points out, only a stakeholder's expectations can (PMI, 2013b, p. 45).

Within the Program Stakeholder Engagement Domain the three activities are as follows:

- Program Stakeholder Identification
- Stakeholder Engagement Planning
- Stakeholder Engagement (PMI, 2013, p. 46)

Throughout the program, the program manager is constantly reviewing the stakeholder list (or register). During the planning phase, the program manager typically interviews the most likely stakeholders (program sponsors, component project sponsors, functional managers, other program managers, project managers, governance board, team members, funding organization, performing organization, customer, subcontractors, and so forth). During this interview, the program manager should always ask who else should I speak to or who else does the program affect? Another key question might be who else might have key information about this program or any of the component activities? This activity should happen periodically throughout the program. Periodically does not mean once or twice a year but would depend on the volatility of the stakeholders and the program.

The interviews should then be mapped to understand the stakeholders' influence on the program, how much monitoring must be done with the stakeholders (PMI, 2013b), and where they may help in areas of complexity. Complexity is about areas where the program is on the edge of chaos.

For example, company X and company B have an established relationship and currently are working on a major RFID upgrade program. Recently, company X instituted and required all its suppliers to use radio frequency ID chips. Think for a moment if company B, a supplier to company X, were using the RFID chip technology and a natural disaster came through and wiped out the infrastructure for the company. Company X has a warehouse intact but no way to ship because the company does not have a means to scan the RFID chips. In order to stay popular in the eyes of public opinion and to keep a good supplier going, theoretically company X should loan or even buy company B at least the RFID scanners to move the supplies in the warehouse. Complexity may present itself in many forms. By mapping the stakeholders and having it available to the program team, this information is readily available if necessary.

Complexity mapping needs to be part of stakeholder planning (Byatt, 2013). How will the stakeholders be communicated with and when are essential questions to be addressed in the planning process; however, as shown in the example above planning at a high level needs to be done for complexity. One trap to avoid is over-planning. Identifying stakeholders with skill sets and how they may help the program is all that may be needed, for example, for a global program identifying internal and external stakeholders, where they are located (Europe, Asia, North America, Africa), expertise, and the internal role on program and level within the company. This information would be useful when a program team might be isolated, such as network outage for a section of Europe. The program team in the United States, Asia, or Africa in most cases would not be able to help. Depending where the program manager is located, he/she may be of assistance. In many cases, the program manager should not step into these cases of complexity. The program manager should have provided the teams with the tools to solve the situations. One such tool is the mapping of stakeholders, and another is the stakeholder engagement plan. This network outage in Europe should be dealt with by the program team in Europe, possibly with the assistance of stakeholders in Europe. These stakeholders may be the IT functional manager, governance stakeholder, outside IT vendors, and others. The types of stakeholders will be highly dependent on the situation, type of program, and type of stakeholders.

Stakeholder engagement is a critical element to the success of a program. The program manager is continuously testing soft skills when engaging program stakeholders. As with any program, stakeholders may or may not have the best interest of the program. The stakeholder may be the component project manager sponsor. Program managers may accelerate and slow or stop component projects for the benefit of the program. This may not go over well with the component project manager sponsor and creates a complex situation with his/her organization, including budget, schedule, and even personnel issues.

The program manager in these instances needs to ensure that the affected stakeholder understands the goals of the program, working with the stakeholder to re-deploy resources, to accurately reflect risks to the stakeholder's organization, and to provide accurate and honest communications. When stakeholders perceive the communication is open and honest, done well, and the good of the program and ultimately the company is at the heart of the plan, most stakeholders will support the needs of the program. The support may not happen immediately and may even take the program manager several weeks of negotiation and engaging the stakeholder.

In a complex environment, the messaging needs to be simple: there should be an ownership (logo/phrase), the objectives/goals need to be understood by everyone on the program, the message must be believable, and stories should be used for messaging (Byatt, 2013; Gordon & Curlee, 2011). A program is a culture, and the program manager needs to instill a sub-culture of complexity. Culture is a system of shared assumptions. Assumptions help guide the program team toward resolving internal and external challenges. The culture provides a framework for the program to interact, judge, and even intimidate external organizations and forces. Understanding a foreign culture allows an individual a paradigm to interact with it (Schein, 2004). Cultural complexity is the understanding that shared assumptions offer a means to judge guidelines and principles that define and guide a group (Curlee & Gordon, 2010). The framework is not formal or hierarchical but is one developed through leadership, experience, observation, and organizational folklore.

Shaping culture would appear to be a missing component in addressing complexity theory in dealing with the many stakeholders. The program manager cannot be present to address every challenge or deal with every stakeholder (Krajewski & Ritzman, 2001). The program manager must have the vision of what the program needs to look like to mesh with the organization and succeed with the stakeholders. Culture takes time to shape, and the program manager does not impart the culture overnight. Merging culture with vision needs to occur to allow the program to move forward in unison. The component projects need to adopt the vision as well as the program. The culture and vision are occurring from the top down and the bottom up.

A complexity-based leader cannot be limited in expressing ideas and concepts. Successful complexity-based leaders when establishing cultures on their programs use myths and metaphors. By combining with icons, symbols, slogans, and other codes, the leader can drive the program team to above-normal achievements. The complexity culture drives the program to a more social culture.

Myths and metaphors may be used to leverage a team and even stakeholders toward a higher level of change. They are expressions of the complexity of human systems. Myths and metaphors have been known to move teams and programs toward greater success. Great program managers may use an organizational myth as a central theme for a program to transform it beyond the actual needs of the program. Think of 9/11, the BP oil disaster, and the natural disasters in Japan. The leaders of these programs did not have to go far to design organizational myths to have the nation pull together to help. These organizational myths bring about a social phenomenon that is as great as the *butterfly* example in complexity. The flapping of a single butterfly's wings (or actions of a single person) can create a great change or effects elsewhere or later in time. Think of all the flights that were safely landed in a record amount of time in the United States on 9/11.

Most organizations have stories to tell. This is the making of the myth. The program manager must craft the details and make sure the important elements are retained and the message is maintained. The program manager can take a pertinent program milestone, how an individual was effective, or how the complexity was overcome, and talk about the obstacle with a few team members. Good news travels just as fast as bad news. Create myths about people on the team and about the program itself. Soon the stakeholders will be supporting the program, even when there are hard times. Remember that people are social creatures, so there is nothing wrong with creating positive myths about good people on the program.

A military officer learns early in his/her career to present several options to the senior officer. The junior officer also should present his/her recommended option. The senior officer may have two or three junior officers present the same scenario and not every junior officer will provide the same recommendation. Is only one officer correct? Possibly, if the scenario presented is a linear-type scenario. On the contrary, if the scenario is complex, the answer most likely is they were all correct. Why? It would depend on the rationale, assumptions, and the complexity at the time. The next time it happened all the same officers may present entirely different answers. Complex problems do not have only one right way to succeed. Too often leaders, hence stakeholders, become set in how it was done in the past rather than in looking how it could be done successfully in the future. There may be elements that are similar to past situations, so the program manager should look for those (and other leaders on the program as well), but those similarities should not be the primary course of action.

Stakeholders with experience may attempt to replicate their experience in every new situation. Rather than learn what could be done, too often individuals do what has worked in the past. The program manager should help stakeholders on the program (e.g., component projects, component activities, component programs) to learn and then apply a solution, and then the program is ruled by complexity. If there are pressures upon the program leaders to seek the best solution rather than apply a new solution, then the organization is truly embracing complexity. A culture of complexity should supersede the individual, and the culture should be driving change.

Program managers must help the stakeholders realize that everything is connected on a program. This is related to the butterfly effect as mentioned earlier. Benefits realization is one of the program fundamentals. A minor change to the company's environment or the program's internal environment, or even a shift in the industry (the flapping of the butterfly's wings) may cause a ripple effect that forces the program manager to stop some component projects. Complexity represents elements of unknowns. The program manager needs to review and understand what variable caused the change to understand the trend and how it affected the program. The variable may cause changes in other areas of the program not normally expected.

Program managers pay attention to the interdependencies of the component projects and programs. The environment that made the program successful might change in such a way to make a program fail or no longer be necessary. Failure should never be an option for a program manager. Cancellation of the program may be necessary, but it should never be a surprise, especially for the stakeholders.

When creating the program culture, program managers should blend it with their own distinct philosophy. A culture has the similarities of a tribe. Upon understanding this, program managers can drive their ideas into programs and thereby expand their sphere of knowledge. Program managers impart their beliefs through intellectual and non-intellectual means. Business culture in the Western world is based on free enterprise and has a strong basis on survival of the fittest (Curlee & Gordon, 2010).

Complexity-driven leaders understand that their role as a modern leader means they will have limited time and limited contact with everyone on the program. Given that the leader will not always be available for the team and stakeholders, there must be a way that a surrogate leader can be consulted. In the past, this was accomplished by having multiple layers of leadership and the program manager would delegate authority and responsibility to others. This situation is limited and often more expensive than a program can allow. The most important aspect of complexity-driven leadership organizations are the utilization of a clear leadership philosophy that offers direction, guidance, and clarity to questions others might have about the program. Complexity-driven program managers understand the social system of individuals and realize there is not a direct relationship for critical social concepts, such as learning, knowledge, and understanding (Curlee & Gordon, 2010).

Complexity leadership is difficult but intuitive for many program managers. Organizations need to provide the program manager the leeway to effectively utilize complexity. For complexity to be successful in an organization, it requires that there already be a culture and system of solid leadership supported by the culture. Furthermore, the program manager must be an individual who is respected by the program and is considered the correct person for the job. Having a solid footing to start will assist in allowing the program manager to be successful and earn the respect of the various stakeholders.

Politics is endemic in any organization and normally extends to the program. Humans are not machines like cars or computers. People can always choose to make a difference or to be a roadblock, depending upon their political disposition. Widgets in a machine have no such free will and will always perform their assigned duty. The widget will either be 100% or 0%. People simply do not operate the same way. People will offer productivity between 0% and 100%, and this productivity will depend upon a great number of variables. Unlike the widget that has two states, working and productive, or broken and non-productive, people have a wide range of productivity.

The program manager must understand that he or she must learn to navigate the political infrastructure of the organization and the program. There is always an unwritten dynamic among the strategic stakeholders of the company and those of the program. While some individuals will have agendas, either hidden or apparent, the goal of the program manager is to mobilize the stakeholders into action to realize the benefits of the program.

Consider the program as a set of small teams that are interdependent and the program manager is expected to synergize the teams to increase the benefits. Small teams support complexity for communication, accelerated learning, and achievement. The small teams create networks that are helpful later. These are external stakeholders that assist the program in times of complexity to help with the success of this small team and ultimately the program.

Programs have become more complex and will rely upon fewer individuals to produce greater results. Complexity theory simply stated is there are systems too complex to define but appear to have patterns with some meaning. As stated before, the program manager must keep in mind certain previously discussed tangential theories that form the basis of complexity theory, such as the butterfly effect and six degrees of separation (Cooke-Davies, Cicmil, Crawford, & Richardson, 2007).

A successful strategy is for the program manager to calculate the formal and informal lines of communication (LOCs). Understanding how information flows within the program makes for an important exercise for the program manager. Keep in mind that there will always be formal, hierarchical lines of communication and informal, non-hierarchical lines of communication. It is best to map out these lines in order to better understand how information is disseminated within the program. Mapping out these LOCs between the strategic stakeholders helps the program manager

Stakeholder Engagement

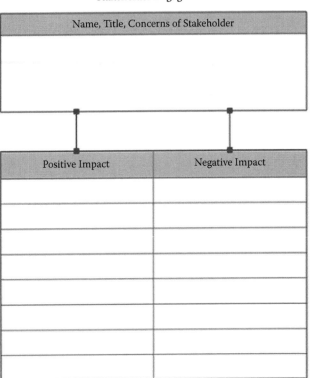

FIGURE 12.1
Stakeholder engagement.

understand the relations among the stakeholders and the complex nature of the program.

As a program manager, consider using the engagement form to identify and map out every relevant stakeholder. This can offer a profile about each person involved and can be used as an area of focus and concern. Understanding stakeholder concerns and identifying them in advance is important for effective stakeholder engagement and communication (Figure 12.1).

13

Successful Program Management and Complexity

PHOTOGRAPH 13.0

A white castle and flowers symbolize the marriage of war and peace.

INTRODUCTION

Programs are an amalgamation of different projects that are connected with a strategic goal. Complex programs require a variety of concepts in order to address the economic issues related to the program. Ultimately, a successful program is one that realizes its benefit and is completed on or below budget. Although time is a consideration, if the target budget is achieved, this is a major achievement for any program. In order to achieve successful economic results in a program, one needs to consider all the life-cycle costs of the program. Most programs consider the costs of the program, but one must always consider the recovery costs associated with any long-term program. To this end, one needs to consider both the forward logistics and the reverse logistics of a program.

Reverse logistics management is about the handling of materials after they have been sold to the end user—in this case, this would be materials that were used for the program but were not fully used or consumed in the process. How this impacts a mega-program is that within any program there will be materials purchased, modified, and used to finally build and complete the program. Given that so much material will be acquired during a program, it is important that attention be paid to the forward logistics as well as the reverse logistics. An analysis of the waste stream and recycling should be completed to address not only any environmental concerns but also potential economic savings.

The three essential areas of reverse logistics are returns, reclaimed goods, and recycling (RL3), and a properly engineered reverse logistics department within a program will allow all of these areas to work together while everyone has visibility of these processes throughout the organization. Visibility is about allowing everyone in the organization to be able to retrieve, review, and analyze the information about reverse logistics. Complete visibility is important because then there can be a true economic analysis regarding returns, reclaimed goods (waste management), and recycling. Information is the first step to achieving better outcomes, because since many of these elements are typically ignored, if one can design an organization that can leverage these areas it can lead to savings throughout the program.

In a mega-program, such as the launching of the Disney Cruise line that included the construction of two large modern cruise ships, the construction of a dedicated terminal in Port Canaveral, a fleet of buses, and the acquisition and development of a private island was a billion dollar

investment (Hemmingway, 1998). If one were to make a conservative estimate and consider that 1% of a billion dollar program consists of waste
and recyclables, then a billion dollar program has a $10 million opportu-
nity, if managed properly.

If one were to start in the planning phase of a program to consider this
opportunity and to develop systems and management that would maximize these kinds of opportunities, it could certainly offer considerable
program savings. Given the current economic situation, no business can
overlook this kind of potential and so it must plan for addressing these
economic and environmental realities.

Returns Best Practice

The reported value of U.S. returns is estimated at $100 billion per year
and consists of approximately 4% of the U.S. gross domestic product
(GDP) (Blanchard, 2012; Li & Olorunniwo, 2008; Stock & Mulki, 2009).
This statistic alone shows the importance of the management of returns.
Further studies have shown that the rate of returns can vary between
5 and 50% (Rogers & Tibben-Lembke, 1998), and so even at a modest 5%
rate, this level of returns is significant. Given this level of potential, any
mega-program should start with a manageable process to address returns.
What often happens is that returns have many more processes or steps and
so are less effective than traditional forward logistics which leads to more
money lost through inefficiency.

The returns process is the most commonly identified aspect of reverse
logistics. This is not surprising, because all retail organizations need to
have some manner of returns process. What was found in much of the
reverse logistics research was the organizations that did not have a solid
defined process for their returns had the most room for improvement
(Blanchard, 2012). If the returns process is not done properly, an organization can lose the value of these returns. A best practice in returns in
a program is to use a tracking number, such as a purchase order is used
for supply chain and forward logistics, and this internal return number
allows the return to be better tracked (Gordon, 2011). The consistent use of
this kind of tracking number was found to help improve the returns process as it moved through the organization. Keep in mind that if a product
is not returned, the loss is 100%. Incorrect, damaged, or otherwise unusable product can be a drain upon a program if not managed properly.

Another best practice is to appoint a person to oversee the reverse logistics process within a program. The establishment of the reverse logistics department is an important way to make sure that material recovery is part of the program. Depending upon the size of the program, it might not mandate a full-time position, but making someone responsible and having someone skilled in negotiations will certainly assist with savings in the area of returns. Organizations have long understood that having a highly trained and professional supply chain group will yield organizational benefits beyond the cost of these professionals, and organizations are only beginning to realize that applying the same level of training and professionalism to reverse logistics can yield the same level (if not greater) of benefits. Furthermore, another best practice in returns is the policy and customer service. Given that no process will be 100%, a manager in a program must understand that there will be incorrect, broken, or otherwise unusable materials that will end up on the loading dock. Since the mega-program will have to manage hundreds or thousands of suppliers, it becomes important to have a reasonable returns and customer service process. Without having this requirement to start, the program might find itself overspending just to acquire the necessary materials to complete the program.

Reclaimed Goods Best Practice

To clarify, since there are many competing ideas in this area, recycling in this context is any program that reuses material by returning to locations that reuse material in a similar fashion. Examples are cardboard recycling, aluminum recycling, and plastic recycling. These materials are recycled and reused in a manner that was similar to the original item. Some of these materials might also include a bounty for returning the items for recycling (such as CA redemption value) which is a cost taken at the time of the sale and is returned if the product is returned to an approved recycling center. Reclaimed goods are materials that are either re-sold in a used state to gain back some value, such as selling old computers or materials that are sold for scrap, such as the sale of old cell phones where if the model is too old the item is reduced to usable pieces or even to the point of extracting certain rare trace elements used in such technology. Both of these areas are important in a program, but each could be handled differently, depending upon the program or the volume of the particular areas.

The management of the waste stream has become more important as the public has become more aware of organizations that do and do not take steps toward preserving the environment. There are economic issues that should be pushing program managers of mega-programs to make sure that all of their stakeholders are environmentally conscious but also fiscally conscious of waste. The United States is by far the most wasteful nation in the world, and although we have taken steps to reduce waste and to recycle more, the United States is still far behind other nations.

To understand the potential untapped recycling potential in the United States, if we take a state like Ohio that has a strong recycling industry, it can offer potential to a program. The Ohio Department of Natural Resources reports that Ohio's recycling industry generates $22.5 billion in direct sales and employs more than 100,000 people. Recycling in Ohio accounts for $650.6 million in state tax revenues, and the numbers presented make it clear that Ohio has been successful in tapping the potential of recycling. Ohio is a leader in both employment and sales derived from the recycling industry. Furthermore, to understand the scale of the success, 4.3% of the workforce is involved with recycling. This percentage far exceeds neighboring states of Indiana, Pennsylvania, and New York. The average annual salary earned by Ohio recycling workers is $36,600. This average salary is higher than comparable jobs in nearby states despite the cost of living being higher in New York.

Despite political and social pressures, some companies still actively resist moving toward programs and solutions that preserve the environment. Although there is no clear solution for every program, there are certainly three important steps that any organization should take to make sure that they are addressing and potentially saving money through recycling and reclaiming goods. There are three best practices that stand out with regard to recycling or reclaiming material.

The first best practice is to appoint a responsible person who has a passion for this kind of work. If a program manager does nothing else in this area, appointing someone accountable will at least help move the organization in the right direction. It is not necessary for a top-level manager to be appointed the environmental manager, but what is important is that the person believes that it is important and is passionate about making sure that it gets done. Too often organizations are satisfied with putting colored bins out and leaving the responsibility to the employees to properly sort. Without any kind of accountability, these kinds of programs never amount to much and probably end up costing the company more money than they save. Making a passionate person responsible will often lead to remarkable results.

The second best practice is to require that all areas have a recycling and reclaimed goods program in place that is sound and sustainable. When a program is required it forces people to think about the best way to do it, and if the group is forced to bear the economic impact they will try to make the program as cost effective as possible. Too often a program is set up at too high a level and does not take into account the correct information. A program will have resources on a global level, and a recycling program in New York City might not be the same approach to take as a recycling program in Tokyo. One also needs to take into account the scope of the recycling or reclaimed goods; if the volume is high then one should consider utilizing a liquidation company that would be able to assist with locating buyers for the recycled or reclaimed goods (Rogers & Tibbens-Lembke, 1998).

The third best practice is to examine the carbon footprint of the program. This might seem a lofty goal because there is no requirement for such a review, but if a mega-program can incorporate such awareness, then it will help encourage more awareness of the environment and the mega-program. Depending upon the program, there could be opportunities with energy efficient buildings, energy efficient technology, and the use of alternative energy sources to attain a carbon zero footprint (Gordon, 2011).

SUMMARY OF REVERSE LOGISTICS AND PROGRAM MANAGEMENT

Reverse logistics management is important to a program because it can better utilize the acquired resources for a program. Since a program will typically span years, a longer time horizon must be considered. The more time that the program will take, the more resources will be used and the more resources will be obsolete by the end of the program. By considering these long-term issues such as returns, recycling, and reclaimed goods, the more efficient the mega-program will be. In one building in London that was built in 1870, through effective management of resources, leveraging alternative energy sources such as solar and the harvesting of rainwater, the building was able to achieve through renovations a 60% reduction in energy and water requirements (Gordon, 2011). Such optimization shows that construction of a new building or development of new technology is not required for an organization to achieve remarkable savings.

Programs have significant opportunities in so many areas. There are multiple best practices that can achieve significant savings in scope, time, and budget. Any program manager tasked with being involved with a mega-program can apply many of these best practices in order to improve the area or perhaps even the entire mega-program. Opportunities exist not only in process and people but also in programs. Process improvements exist by applying the best practices to programs, while unrealized savings can be found through efficient recycling or reclaiming goods that otherwise would have been wasted. Given that programs will span longer periods of time, it is important to understand what can be achieved and to understand that economic impact and savings can mean the difference between success and failure of a program.

PHOTOGRAPH 13.1
There is still so much to be written about program management. The field is wide open!

14

Course Materials #4

PHOTOGRAPH 14.0

There is a narrow road between success and failure. The case study presents a difficult situation where a program manager can show that he or she has the ability to navigate even the most treacherous paths in life.

CLASSROOM MATERIALS

Case Study #4

Military Technologies Inc. vs. Guidance Systems LLC—Part Two

Re-read and review the case study in Chapter 10 as a point of reference for this case study. You are the program manager for Military Technologies Inc. and Guidance Systems LLC is currently the subcontractor to Military Technologies Inc. for other guidance systems for the new generation of military support ground vehicles. These vehicles are state of the art and have considerable technology on board to move the vehicle as well as to notify the driver of friends and foes. This vehicle is under development for ultimate use by the U.S. military.

The negotiations that were discussed in Chapter 10 have already taken place, and this case study is taking place about a week after that negotiation meeting. From that meeting, a verbal agreement has already been discussed that sounded like it would meet everyone's expectations. Although the agreement to utilize the new proprietary technology between Military Technologies Inc. and Guidance Systems LLC has not been signed, most of the agreement was concluded in the last negotiation. Jack Smith was given the responsibility to complete the contract details, but then without much notice, Jack Smith resigned from Military Technologies Inc. and took a job with a competitor. Jack Smith did not have a confidentiality agreement and did not have a non-compete clause in his contract, and this new situation could potentially be an issue.

What is interesting is that Jack Smith has taken a new role with one of the companies that had been put forward as a potential alternative supplier by Graham Garnet. Graham feels that this was a planned move by Jack Smith and feels that Jack can use the information from the negotiation as well as from the file that Graham had compiled to destroy all the hard work that has gone into the negotiation. You are not sure what Jack might be up to; however, you wish him well.

As you are working to complete the contract between your company and Guidance Systems LLC, you get a call from Jack Smith. Jack lets you know that his new company, Seatech Industrial Inc., can actually provide the same technology at half the price of what the deal with Guidance Systems LLC was going to be. He advises you that now that he works at Seatech Industrial, he has reviewed their specifications in detail, and he is sure that Seatech could do the same thing for a lot less money. You are unsure of what to do because the savings would represent a lot of money to your company.

Later that day, you get a call from the head of Guidance Systems LLC, Edward Saturday, and he advises that they got a call from Jack Smith that you are considering an alternative offer from Seatech Industrial Inc. Mr. Saturday tells you that he feels upset that at this late stage in the negotiation you would consider an alternative offer, from an inferior company. Edward Saturday speaks about the long history between the two companies, and how the offer that you have made to Guidance Systems LLC is a lot of money for a small company, and it means a lot of new jobs. Backing out now would not only mean no new jobs, but it would result in putting dozens of people out of work. He mentions to you that he knows that the U.S. military has been promised this new proprietary technology from Guidance Systems LLC, and he would personally go to his contacts there to tell them about how you purposefully misled everyone involved.

Mr. Saturday advises that you may have worked with Jack Smith for a long time, but he says that working with him was not easy. He found Jack to be deceptive at times in the relationship between the two companies. He would hope that you would do the right thing and send over the contract at the agreed upon price. You close the discussion that you will call him in the morning after you have had some time to discuss the matter internally. Mr. Saturday was not happy but advised that he will look forward to your call.

Case Study Questions

1. What do you do? Will you consider the new offer or will you move forward with the contract as verbally agreed with Guidance Systems LLC?
2. Do you use this opportunity to re-open negotiations with Guidance Systems LLC? It is obvious that they are worried, so do you use this as a way to get a better price?
3. You feel that given the politics involved in this now, you will likely get a call from the president of your company to ask about the status of the contract. What will you prepare as a response to the company president?
4. What will you cover in your call to Edward Saturday at Guidance Systems LLC in the morning? You know that you need a solid position with whatever you decide to do.

5. Will you include Graham in this discussion? Why or why not?

6. Do you contact Jack Smith? Why or why not?

SECTION QUIZ

Section 1

Multiple Choice

1. Words account for a small part of communication. What are two other aspects that account for communication?
   a. Tone and the way the person is dressed
   b. Body language and the way the person is dressed
   c. Demeanor and the way the person is dressed
   d. Tone and demeanor

2. Technology and several other factors help the virtual organization. Name two others.
   a. Training and processes
   b. Training and face-to-face meetings
   c. Processes and face-to-face meetings
   d. Organizational meetings and face-to-face meetings

3. English is considered the international business language. Which one of the following is a barrier of the English language?
   a. English's nuances
   b. English's various written forms
   c. English slang
   d. All of the above

Section 2

True/False

1. Modern virtual organizations can only exist with technology.
   a. True
   b. False

2. Program managers have a strong, stable relation with their direct supervisor and this helps with the stability of the program.

    a. True

    b. False

3. Always use the latest technology for the program.

    a. True

    b. False

4. An employee may suffer from loneliness when shifting from a traditional organization to a virtual organization.

    a. True

    b. False

5. LinkedIn' is an example of social media.

    a. True

    b. False

Section 3

Answer Key

Section 1

    1. D

    2. A

    3. D

Section 2

    1. A

    2. B

    3. B

    4. A

    5. A

DISCUSSION QUESTIONS

Short Discussion Questions

Technology can help a virtual organization succeed or fail. What are some of the areas that a virtual organization may overlook when it comes to technology?

As a program manager, discuss how you would harness the power of social media for your program.

PHOTOGRAPH 14.1
Everyone admires success.

15

The Future of Program Management and Complexity

PHOTOGRAPH 15.0

Modern technology vs. nature's tranquility. Birds have adorned the skies longer than jets. Birds instinctively know where to migrate; humans use technology to migrate for business, to relax, and to move.

COMPLEXITY THEORY AND COMMUNICATIONS

Communication is instrumental for the success of any program. Intertwined with this success is understanding how to use communication effectively in a complex environment. Adapting communication leadership skills to complexity theory and arming the program team with tools and techniques to take advantage when faced with complexity will lead to teams that are ready to tackle the unexpected. To achieve this goal, the program manager must learn to leverage communication technology, maximize face-to-face contact, and send important messages in multiple formats to ensure reception by all stakeholders.

Communication must not be limited to any one media; multiple media formats are preferrable to ensure that the message is communicated. This redundancy of communication is not easy to accomplish, but it should be applied whenever possible. Program managers must understand that component projects will span many different groups and that additional communication is necessary. An important factor along with this communication is accountability of all stakeholders involved. Communication must not only be clear on expectations, it must be clear on consequences. Program managers must be even more diligent about holding parties accountable to their responsibilities (schedule, quality, statement of work), because the programs that they manage might not be seen as the largest priority at the time. The day-to-day program tasks might consume the component project manager, but from an organizational goal or success standpoint the success of the program is a direct reflection upon the organizational image. Successful programs benefit the organization, and the more programs that become successful, the more people will want to participate.

Communication at multiple levels and through multiple media can assist a program to become successful. One successful program can assist a program manager in gaining additional support in other programs and projects. For example, if a recycling program is successful, other project managers will seek to be involved on component projects in order to share in the organizational success. The only way that people can know about a successful program is through good organizational communication. Communication is more about passing along information to help shape the opinions of others than about forcing compliance from unwilling parties.

Gordon Moore, the co-founder of Intel, astutely predicted that computing power would double every two years (1965). In his paper (1965), he also stated that integrated circuits would double in capacity every 18 months. There were other accurate forecasts about technology that add to the credibility of the paper, and Moore's predictions became known as Moore's law. Overall, he was correct.

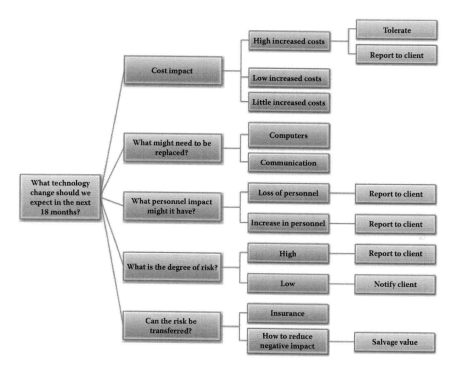

FIGURE 15.1
Moore's law.

Assess potential Moore's laws issues by examining different technologies in the program and use a flowchart similar to that shown in Figure 15.1. The assessment should review risks and costs.

Whether the advances made in technology will be those of Star Wars or Star Trek, holographic images and robots with feelings speaking with humans, or communicators translating on the spot, only the future will dictate, but there is no doubt that technology will advance program communications. Technology has helped program managers implement complexity theory on their programs.

Social media is based on the complexity sub-theory of six degrees of separation. Without the bright minds of college students meeting the needs of how to socialize and share information, social media would not have been born, or at least not in the form we have today. Business may not always be as creative as college students meeting a need for themselves and their fellow students.

Consider cell phones and smart cards. Both technologies were created by the same phone company that had no idea how to market it. Both were abandoned, and the ideas were given to the open market. History shows what enterprising, out-of-the-box thinkers can use with technology that is not yet used. Where would you be today without your smartphone?

Enterprises have business plans, competing funds, and milestones that must be met for new technologies to be created. Creative program managers adapt or push the limits of existing technologies to meet program needs. Global programs are faced with some of the following technological issues which hamper complexity:

- Language barriers
- Lack of virtual reality/simulation
- Time zones/geography
- Dependence on hardware (e.g., must stay linked to one device for video conferencing)
- Bandwidth
- Email volume

Other issues exist, and each program manager, depending on his or her situation, can add or subtract from this list. Each program is unique but will have commonality with other programs. The above matters will be evaluated for emerging technologies.

Language barrier continues to be an ongoing challenge on programs. There are several official English languages (United States, Australian, Great Britain, etc.), and the nuances between them are many. There are also many nuances within the English language. This creates difficulty for the individual who is not a native speaker. The program manager also must establish a program communications plan that establishes clear guidelines on the use of the English language in emails, during conversations, and so forth. This can stifle creativity during complex events if individuals are not using their native language.

A new technology being tested is *augmented reality*. This would be an app that resides on a smartphone or in special glasses. It would allow the user to interact with an individual in real time while each is speaking his or her own language. Each person would need to have the app loaded on the special glasses or smartphone. Some businesses already require their employees to upload certain apps onto their phones, so it would not be a stretch for companies to require this particular app.

Augmented reality may be taken a step further, for a virtual water cooler break. This would help prevent the loneliness that often develops for those who first transition onto a virtual program/project. Think of the potential for those incidents on the edge of chaos. Small teams could actually meet in a lab, where the disaster occurred, at the construction site, or in a conference room, and each person would speak in his or her native tongue to resolve the crisis at hand. Think of the possibilities this technology has for those who might be able to adapt it for other program uses. Training could be done globally, and resources may not have to be redeployed.

In the 1990s, video teleconferencing was one of the premier technologies. Companies that could afford the technology were proud to demonstrate its use. The technology was difficult to use, employees had to be very still during the conference, sometimes they had to travel to a central location, and many times it did not work or was not compatible with another company's system. Video teleconferencing soon fell out of favor by many companies as expensive and not worth the trouble.

In the late 1990s and 2000s, PC webcams became the next technology wonder, and many laptops come equipped with them installed. Program managers may use webcams on occasion to host video conference calls to add a personal touch and help the team members get to know one another. The issue with the webcam is that the person is also tied to the laptop. As tablets and some smartphones provide the ability to have face-to-face chats, then we are no longer tied to a laptop to have a video chat. This lack of a tether to a laptop would allow the employee to roam to the location needed to have the video chat to resolve the complex situation. One person from the team may be at home where it is 2 a.m. and he needs to move to the study so as not to wake the family; another member is at work in a lab but moves to a conference room to have some silence; and the third person is the team lead. This would be even better if this team was using the augmented reality app, especially if there were a language barrier and if they were resolving an *edge of chaos* situation.

Program managers can become overwhelmed with emails quickly and fear that they may overlook a critical email from the client, program sponsor, or team member. Many program managers try to find an hour or two of quiet time each day to quickly go through the vast amounts of email, but in reality most emails are just scanned. Program managers in the future may be able to directly link to a computer. So during those sleepless nights, when the brain had that brilliant solution for the perplexing problem on the program, or the brilliant answer for the client, the thoughts would telepathically transmit to the computer. This technology may not cut down on the number of emails that need to be read, but it will capture the creative thoughts that we lose in our sleepless or drifting off to sleep moments.

Social media sites are addressed in this chapter. Without social media sites, complexity theory might have stayed in the area of mathematics. Social media has skyrocketed and has been accepted by businesses. Those businesses that understood how to harness the power of the social media sites have been able to extend their reach officially and unofficially. Officially they have expanded by having a company site. The company also receives tangible marketing and advertising with each employee who has a profile with a resume containing the name of the company as a current or previous employer. Employees also may tweet about the company. This may be at the request of the company, or the employee may be responding to some event. In either case, the company is receiving publicity (it may be positive or negative).

Internet of things (IoT) is another interesting technology that will aid the program managers as teams become more adept with smartphones and IoT becomes more readily available (Grier, 2013). IoT is the ability to manage "things" such as making a cup of coffee, accessing data from the laptop, and so forth from a smartphone. This technology is just starting to become evident in the marketplace. There is a car advertisement where a spouse wows her husband by locking the car remotely while boarding a plane. There is another commercial where a parent activates the home security system remotely using the smartphone.

Eventually this technology may allow program managers who are leading disaster recoveries to move rubble that is trapping survivors, or at least to send a locator beacon. Or this technology may be able to network and analyze the various structures that have damage. The analysis can then be sent back to the appropriate parties with the next steps or safest steps for cleanup.

IoT would have many uses for complex business applications as well. IoT technology may be able to help the program manager predict the likelihood for the edges of chaos on programs. If this technology is able to do this with some accuracy, then automated tools designed to react to complexity can be incorporated within the IoT and integrated into augmented reality. New technology is only limited by the imagination.

Knowledge is power, but many disciplines and companies have so much knowledge/data, the question becomes how should it be managed and accessed? What should be done with perishable data? Many companies have knowledge management systems, but few of the systems are intuitive to the user, many become dumping grounds for data, information becomes dated, and searched data are not returned in any rhyme or reason. Companies' internal knowledge systems are not Google® or Bing®. The end users become frustrated. Program team users may remember something that could help with a complex solution but go to the company's knowledge system and become frustrated because the information cannot be found. A new technology for harnessing *knowledge* is visualization of data (Grier, 2013). There is promising visualization research that demonstrates essential information and data analysis is found much quicker and is more accurate than in traditional methods. For large, complex programs, extracting and analyzing current and historical information is essential. For most companies, this is time consuming and laborious. Visualization promises to decrease the labor and time needed to find and analyze the data.

No one would dispute the power of social media. Throughout the book it has been noted that social media is based on the complexity sub-theory six degrees of separation. As previously explained, any person in the world is no more than six people away from being connected to any other person in the world. Such a powerful social medium should be considered an essential tool for all programs. Yet a McKinsey Global Institute research report (Chui et al., 2012) found that companies vastly underuse the capabilities of collaboration and communication available in social media. Social media technologies include the following:

- Blogs
- Wikis
- Discussion forums
- Collaboration sites
- Social networks
- File sharing

McKinsey suggested that to harness the power of social media across the enterprise, a company "must be open to information sharing and create cultures of trust and cooperation" (Chui et al., 2012, p. iii). Social media is trending to smartphones versus laptops. Facebook's SEC filing noted that by 2015 mobile Internet users will overtake wired Internet users (Holmes, 2012). Social media has now taken care of integrating just about all the social mediums. What does this mean?

A Facebook account allows the owner to link the following:

- Email address books
- Tweeter feeds

- Blogs
- YouTube® videos
- Flickr®
- Links to URLs

Think of all this data on individual accounts, and then there are groups, and there are individuals sending messages to each other as well. There is an enormous amount of social data. Some companies have started experimenting with *social media command centers* (Holmes, 2012) to harness the data. GE has used it to speed repairs to the electrical grid, Wall Street to predict stock prices, and Nestlé to try to boost customer sentiment (Holmes, 2012).

Using social media command centers on programs could help the program manager in a variety of ways. First, the program manager could monitor social data external to the program, which would affect the strategic goals of the company/organization and would cascade to the program. Second, the social data may provide alerts to internal company/organization situations that will affect the program which the program manager or someone on the program needs to understand and develop a plan (Grier, 2013). Third, the social data provides alerts about complexity happening within the program. The social data would then be sent to the appropriate leader to deal with the situation as needed.

The McKinsey report (Chui et al., 2012) found that there is a potential to raise productivity of highly trained workers by 20 to 25% by leveraging social technologies correctly. Many people rely on social networks for professional and personal advice. They have never met many of the individuals who they get advice from. McKinsey's report noted that businesses understand that social media drives revenue and is a valuable tool for understanding customers; however, "only 5 percent of all communications and content use in the United States takes place on social networks" (Chui et al., 2012, p. 2)

As social media technologies become more interconnected, it is possible that email will slowly disappear and the smartphone/tablet will completely replace the laptop. Since communications will be documented on the social media site, there is no need to worry about a record of the data. A company that keeps a social network site and encourages its employees to use it to its full potential will have created security and backup routines. Employees will be communicating instantly and finding information on the social network instead of on the company's outdated knowledge sharing site. IBM has an internal social media site (Chui et al., 2012). One

IBM executive decreased his email volume by 98% by using IBM's various social media avenues (Chui et al., 2012).

Eventually, by using a mobile app for the social networking site, the use of laptops will most likely become something of the past. Some people speak to their smartphones now to make or answer a call or to text or chat. In the future, speaking may be an option for social networking as well.

Programs normally function at the enterprise level. The McKinsey report (Chui et al., 2012) saw social technologies adding value across the enterprise by improving intra- or interorganizational collaboration and communication and matching talent to tasks. By increasing the communication and collaboration through social media, McKinsey (Chui et al., 2012) estimates this would reduce email by 25% and searching time by 35%, which should increase productivity for other tasks by 13 to 14%. The other advantage is that communication via social media is now searchable, whereas on email it was not. The communications and collaboration on social media will quickly define someone's expertise and match the resource to the task. This most likely can be done through communications posted on social media sites.

Program managers struggle with the program communications plan and who should receive what communication. Social media can target the correct *program* audience. This is a variance on the marketing communication, but it can be tailored and can be highly effective. This communication may also be bundled with social data from the social media command centers for highly effective and targeted messages to stakeholders or the client.

Many of the technologies presented above are not new, but we continuously learn new ways to use them. Some technologies such as social media need to be used in different ways, such as in lieu of email. Social media may also be able to create tools to analyze the vast amount of social data generated within the media, which then leads to targeting communications to a specific stakeholder group within the program.

Implementing new tools does not resolve bad practices or create new cultures. Most companies that now use social media did so from the bottom up. The younger generation coming into the company somewhat forced the company to adopt the practice of using social media. It may have started with instant messenger and then slowly went to wikis, blogs, social media sites, and Twitter. More progressive companies may even have Facebook sites.

Even these more progressive companies have not fully harnessed the full power of social media. Social media is a relatively new set of technologies that allow for open communication and collaboration. This openness allows for searchable information since all communication is in an open forum and not locked in individuals' emails. For programs working in an open collaborative company, a piece of information needed during a crisis would not be locked on email, the information would be available on the company's social media. Open and collaborative companies would have all communications, especially for programs/projects, carried out on collaborative sites, with an appropriate search engine.

These companies with open communication cultures do not just happen because the social media technologies are put into operation. Leadership must endorse it and use it. In other words, leadership must *walk the talk*. Processes and procedures must be changed to address social media, training must be provided and updated as the social media changes, incentives must be provided, and there must be something gained by the employee. Some companies have established communities of practice that help to resolve difficult problems/issues, provide best practices, hold discussion forums, answer questions, and post content. Companies that ensure these sites remain lively by having a community manager are successful.

An organizational culture shift nominally takes three to five years to be fully implemented. There are ways to make the adjustment quicker. Leadership may choose to let go of middle managers who do not actively endorse the new culture and replace them with those who do. Senior managers may also be replaced. There are downsides to this approach as the organization is going through a radical change, and changing middle management and senior leadership can substantially decrease morale.

With the increased collaboration and communication, there are risks that a company may face with new technologies, especially in social media. Cyber security needs to be at the forefront whenever information technology is involved. The company's data must be secure and the employees' data must not be able to be compromised. During training, employees should be warned about handling of the company's intellectual property, and about proper behavior on social media. In the future, more companies will be moving away from email and running their programs on social media. For the protection of all, for the increased success of the program, and for increased productivity, the rules of engagement need to be understood.

COMPLEXITY THEORY AND PROGRAM MANAGEMENT

Programs are done in a matrixed environment which adds to the complexity of the program, as there is a lack of visual cues during communication. Communication becomes more important to the success of the program. Eighty percent of communication relies on face-to-face interaction; however, communication is shifting to social media and other forms of electronic media which adds to the complexity of a program. Therefore, the program manager needs to adapt communication leadership skills to complexity theory and equip the team appropriately with tools and soft skills. The program manager should be the mentor and advisor to the team but not the problem solver.

Social media needs to be incorporated as part of the program's communication. Social media sites take advantage of the complexity sub theory, six degrees of separation. Six degrees of separation means that any one person is not separated from any other person by more than six connections. Theoretically, a person who has no connections with President Obama would be able to meet the president within six "connections." Many companies have adopted social media but do not completely understand the power that LinkedIn, Facebook, Twitter, or other sites can provide their

PHOTOGRAPH 15.1
Complexity is the narrow road between applying ideas from the natural world and human constructs.

companies. Program managers should adopt social media as part of the program to help the teams and individuals solve issues and problems that lie on the realm of complexity.

Complexity theory for a long time was a game for mathematicians. Then Edward Lorenz, a meteorologst, was able to graphically display that there is order in chaos, or butterfly wings (Wheatley, 1999). His now famous Lorenz attractor graphic which graphs atmospheric disturbances consistently maps owl eyes or butterfly wings. The graphic was published in the article "Predictability: Does the Flap of a Butterfly's Wings in Brazil Set Off a Tornado in Texas?" that he presented in 1972 to the American Association for the Advancement of Science. For many years, the paper was overlooked because Lorenz was a meteorologist, and hard-core scientists (math and physics) did not review this paper.

Program management has been around since civilization began. There is evidence of program management with the Incas, the Egyptians, and in the Bible (Curlee & Gordon, 2010; Levin, 2012). The Project Management Institute (PMI®) first published a Standard for Program Management in 2006. In 2007, PMI recognized the first 31 program managers to attain a Program Management Professional (PgMP®) status. PMI saw the increasing number of programs in organizations and responded to the market.

Levin aptly noted in the preface of her edited book that "as the profession has been formally recognized, more books and peer reviewed journal articles in the field have been published . . . However, much remains to be learned and shared about program management" (2012, p. xii). There is so much that practitioners want to share, and independent studies need to be done by academia.

Complexity has just made the leap into the business world. Some people from the world of mathematics would claim the business world is using terms incorrectly and that complexity does not truly fit in the realm of business. People in project management at times do use terms from other industries loosely.

Program/project management borrowed from the financial industry when naming portfolio management. This became confusing, and within the program/project management domain there are now two terms *program/project portfolio management* and *portfolio management*, both meaning the same thing. When speaking to a non-project audience, it is important to distinguish the types of portfolio one is discussing.

As complexity starts to find itself within the realm of program management, the project management discipline needs to stay true to complexity theory by ensuring the terms are understood and adapted correctly into program management. When developing terms, they should be unique to the program management field. This will minimize confusion for those outside of the field, and especially those in the social sciences who may want to study program management and become confused with dual terminology.

Complexity theory and program management are relatively new in the world of business. Complexity, as mentioned before, has just recently been recognized in the business world, especially for non-linear thinking aspects, and program management has been recognized as a field in project management. PMI has committed to keep the Program Standard current by publishing a new edition approximately every three years. The Program Standard was published in 2006, 2009, and the most recent in 2013.

PMI sponsors research in the field of project management to further the body of knowledge. The monographs published by PMI and available for download by any PMI member in the areas of program management and/or complexity are the following:

- Situational Sponsorship of Project and Programs: An Empirical Review (2008), Lynn Crawford, Kay Remington, Terry Cooke-Davies, Brian Hobbs, and Les Labuschagne
- Aspects of Complexity: Managing Projects in a Complex World (2011), Terry Cooke-Davies
- Early Warning Signs in Complex Projects (2010), Ole Johnny Klakegg, Terry Williams, Derek Walker, Bjorn Andersen, and Ole Morten Magnussen
- Exploring the Complexity of Projects: Implications of Complexity Theory for Project Management Practice (2009), Svetlana Cicmil, Terry Cooke-Davies, Lynn Crawford, and Kurt Richardson

PMI's current research is as follows:

- Global Perspectives on Project, Program, and Portfolio Management in Government, Young Hoon Kwak
- Rethinking Project and Program Stakeholder Management, Martina Huemann

As program managers and advanced practitioners, understanding the latest research is important in developing a person's career. Research provides clues as to how the discipline is molding to the environment: Are there new theories to explain the changes that are happening? Was that gut feeling just you, or is it a trend that research is seeing as well?

A search of PMI's knowledge center on the complex* yielded 1,650 results. This is a combination of PMI's Marketplace (bookstore) (652 hits), PMI.org (245 hits), the various communities (671 hits), and articles and papers published by PMI (PMJ, PM Network, congresses, etc.) (605 hits). The discrepancy of the breakdown is expected as PMI sells articles and papers from PMJ, PM Network, and congresses in the bookstore and this would account for duplicate count.

Since the word *complex* may include hits for projects that are difficult, a further search was done for "complexity theory," and the total results were 48. The categories can be further divided into complexity theory and communications management for which there were a total of seven hits and complexity theory and program management for which there were 15 hits.

A search for "social media," a part of complexity, on PMI's knowledge center returned only 154 hits. The results were almost exclusively for project management with just a mention of program management in one article. The Academy of Management was reviewed as well. Searching social media in quotation marks did not receive any results. Deleting the quotation marks returned over 1,000 hits. Quick review of the results was more in the area of social behavior rather than in project/program management.

As expected, when the PMI knowledge center is searched for *program* the results are much larger (5,076). Of those results, 1,265 are articles or papers. Academy of Management and IRNOP are two sites that publish peer-reviewed articles.

The Academy of Management search yielded surprising results. A search of "complexity theory" for the *Academy of Management Journal* had 45 hits. None of the hits was specifically about program or project management, although one article related to teams and another was about stakeholders. There were no articles resulting on a search for "program management." For the "program" search there were articles but none were related to the discipline of program management.

IRNOP has several publications linked to its site. A search of several of the publications had a hit rate of 52 journal articles. This included a search of program/program manager/complexity theory. There were only five articles that were program management and complexity theory.

A review of the ProQuest dissertation database demonstrates an uptick in universities approving doctoral candidates studying in the area of program management. Universities granting practitioner-based doctorates appeared to have more doctoral dissertations in program management. Some of these studies combined complexity theory as part of the thesis.

There are practitioner research companies as well. These are companies such as McKinsey, Forrester, Gartner, and Standish. They are different from academia in that they are funded by industry, so bias plays into their research. However, their research is focused toward business and is written to be understood by the practitioner. These studies are also very expensive to buy. Standish does the infamous Standish Chaos Report which reviews the health of IT projects and why the projects fail. In the future, Standish may consider adding programs to its research list.

A search for complexity theory on Amazon.com in the book section resulted in 18,876 hits. Complexity theory and project management yielded 17 hits, and complexity theory and program management yielded 1 hit. Much is written about complexity theory and its jump into the business world, but there is relatively little about its contribution to project/program management.

Levin (2012) was astute to bring SMEs together to write about various aspects of a program. She knew there was a void, and a hunger for knowledge among program managers. She also realized that program managers did not have time to wait for the years it takes for academics to do research. Levin astutely asked respected and certified (PgMP) program managers to each write a chapter in her book. Each chapter takes the reader through the life cycle of a program, from why programs are good for businesses to why programs should think about sustainability, and everything in between.

The program managers of today are the future of program management. Each of the practicing program managers is laying the groundwork for what the discipline will look like 5, 10, 15, and 50 years from now. PMI's Standard for Program Management and other project management standards are done by a consensus of volunteer experts. These experts drive the standards which then provide the framework for the methodologies that are used by industries.

These same experts must stand up to be the leaders within program management. These leaders need to do the following things to mature program management and increase the body of knowledge:

Drive research funding toward program management
Help universities understand what program management is
Mentor up-and-coming program managers
Drive project management culture into the organization, as appropriate

The future can be influenced by volunteering at the various project management associations, but the most influential is the Project Management Institute. Writing articles and papers for the Congresses, *Project Management Journal*, trade publications, university alumni magazines/newsletters, speaking at local project management chapter events, and many other opportunities will drive the future of program management. Remember the butterfly flapping its wings; all these contributions are small disturbances that will create an avalanche of knowledge to take program management forward in the realm of discovery.

That means that program managers need to be open to new ideas and try new strategies such as complexity theory. At this point, complexity theory is still being tested to see what parts really do work in business and furthermore in project/program management. No one will dispute that social media (a sub-theory of complexity theory) has helped business, but companies have only scratched the potential of social technologies (Chui et al., 2012).

Academics need to have access to practitioners to research theories/hypotheses in the realm of program management. This list may include but would not be limited to program frameworks, leadership, communications, conflict management, interaction between components, stakeholder management, benefit realization and management, program failure versus success, globalization and distributed teams, forming programs, sustainability, and so forth. Researchers/academics in their fields of study find problem sets, wrap an appropriate methodology around it to study the problem, and then the hurdles come.

The researcher is limited by two significant issues: funding and participants. Research is not free but in the scheme of what it provides, the information can be invaluable, as long as the researcher is doing sound studies. Companies have been hesitant to participate. The reasons can vary. Companies may believe that intellectual property may be compromised, liabilities may be at stake, or the company may just be embarrassed to demonstrate failed programs/projects.

Ethical researchers take great pains to aggregate the data so no one company can be identified. In many studies, the companies are not revealed unless the company provides permission. What is provided in the study is only the type of company. This is needed so the reader can understand and extrapolate what areas of study apply to him or her. For instance, the researcher may state the study consisted of 10 companies, three manufacturing, four IT, and three mining. This is quite a diverse set of companies and would provide the researcher with data for the industry, but it may also show if there are trends that are not industry specific.

Academic studies take time because they are grounded in quantitative/qualitative research and are on a shoestring budget. A professor heads the study and may have a colleague or two to help, or more likely a graduate student. These individuals also are conducting the research and teaching university classes; hence the study may take several years to complete. Increase in funding normally will minimize the length of the study.

Practitioners have a difficult time with application of the studies to the real world. There needs to be a compromise. Academia is mired in tradition and many times this needs to be done to maintain unbiased research. An academic study should follow the traditional steps for sound research methodology; however, good researchers with previous practitioner experience have taken the step to show how the research affects the practice or helps the practitioner. This should be done with all studies in the program management field. This will help the practitioner base become more accepting of the research community, and the future may be more collaborative between academia and business.

Social media is about collaboration and communication. Trust is paramount in the social media culture. To promote the future of program management and complexity theory, what better way to do it than with social media. Social technology is already being used by PMI to foster open and honest dialogue. PMI might consider using social media to enhance communications between the academic community and practitioners. Even within PMI there is a chasm between the two. PMI does promote a Research Congress but this is attended mainly by researchers, advanced practitioners who are underwhelmed by the regular Congresses, and by practitioners with terminal degrees.

By overcoming the perpetual academic/practitioner divide, PMI will have taken a major step toward conquering the fear of practitioners to collaborate with academia. Without this collaboration the future and understanding of program management and those theories, including complexity, will never be thoroughly understood. In fact, the discipline may be using them incorrectly and in a misguided fashion.

PMI, companies, and other associations offer their practitioners social media avenues commonly called *communities of practice* (COPs). These can be wonderful avenues to move the program management community forward. These sites tend to work most effectively when moderated by someone who will keep the site lively and will maintain decorum among the participants.

These sites offer participants the opportunity to share ideas, post problems or issues that have been troubling the program, post best practices, post documentation, obtain training, find links to other sites, and so forth. As these sites become more commonplace and individuals become more adept, the possibilities will grow. The one major flaw is that these communities are closed. Companies only let employees join the communities in the company. The Project Management Institute's (PMI) Community of Practices' are only open to PMI members, and this may be extrapolated to other associations. While discussions, disagreements, and even arguments where the moderator must step in occur, these are generally like-minded individuals.

PMI, companies, and associations would benefit by allowing outsiders into the fold. Look at the LinkedIn® model. Anyone can apply to join any group. Some groups allow everyone, while others have a vetting process. The groups that have an active moderator are more vibrant and tend to have active discussions. LinkedIn groups tend to be global and hence diverse, which provides various perspectives on a topic. PMI, companies, and associations should consider actively opening and moderating groups to further the practice of program management and complexity.

Individuals should be vetted into the communities of practice at any company or association. Persons wanting membership into these communities must be willing to provide some limited data about themselves. The company and/or association should be able to have private areas on the community to protect sensitive data or intellectual property. Companies can also provide training to their employees on the rules of engagement for discussions on the public part of the community; however, companies need to think about the fact that with too many rules, trust will start to erode, discussions will be stifled, and the community will die. Associations must think about the same thing with its volunteers.

This openness may seem uncomfortable for leaders at first, but social media helped the BP oil disaster with many of the solutions. Complexity comes from being on the edge of chaos. By presenting situations to the social media group, the discussion may result in the edge of chaos because of the diverse cultures, but the program manager or program team lead as a result will have many different solutions.

All of these communities are ripe for an academic to join and research program management and complexity. Since the communities are public, the academic researcher would be able to join and have a new avenue for research.

The program managers of today need to collaborate via social media with each other, with academics, and with associations to drive program management forward. The practitioners who understand the latest research along with what is being published in trade journals should be the ones to give back to program management. How? They should drive setting the new standards, collaborate with academics to write articles that provide the practitioner's perspective, write trade journal articles, write books, mentor journal personal, and constantly watch for new trends. Critique if it is a fad, but embrace and keep an open mind should a theory have applications to move program management to new horizons.

Now that you have learned more about complexity, you may want to try using the form presented in Figure 15.2 (Figure 1.1 in Chapter 1) to fill out your new personal concept of complexity and how it can apply to

Advanced Complexity

New Advanced Definition	
Linear	**Non Linear**
Static timeline	Dynamic timeline
Resists change	Allows change to enter
Hierarchical communication	Dispersed communication
One way flow of information	Information flows in all directions
Milestones	Targets
Single point of failure	Multiple contingencies
Single person	Teams
Pipes for all flows	Wave action

FIGURE 15.2
Ideas about complexity.

a program. List more details about the elements below the definition, and compare this to when the form was filled out earlier (at the start of the book) and see how your ideas have changed or evolved.

PHOTOGRAPH 15.2
Program management is about harnessing nature with technology. One needs to be able to connect the natural with the virtual.

16

Advancing the Future of Program Management

PHOTOGRAPH 16.0

The protection of pristine beaches became a significant issue during the Gulf spill.

INTEGRATION OF PROGRAM MANAGEMENT, COMPLEXITY, AND COMMUNICATION

Introduction

This chapter addresses program management, complexity, and communication. To explain how all three of these elements can integrate together, the example that will be used is the BP oil spill in the Gulf of Mexico. This massive undertaking that combined so many different people, companies, and resources can only be described as a program. The program's overall goal was to restore the Gulf to its former level while preserving the environment. However, BP clearly had the additional goal of controlling the massive oil well that was a source of revenue.

Initially, BP staff and management addressed the Gulf oil spill in the traditional manner outlined by the company's published spill response plan. When the crisis first occurred, BP moved forward with an organized response as outlined by the plan. However, the plan was underdeveloped to address a situation of this magnitude, and leadership lost control of the actual work streams occurring in the field. The process showed that initially the response seemed to follow a normal, formal risk plan, but ultimately it was improbable that such a system would be successful. Once the magnitude of the spill was realized, BP and other organizations had to mobilize a vast armada of equipment, people, and materials to combat the spill, which was larger and more complex than anything that had been seen before in the United States, and it became necessary to develop new systems and processes in order to be successful in the shortened amount of time necessary to mobilize all resources in an environment that provided for safety, proper communications, and environmental cleanup.

Four critical findings were discovered regarding the BP response to the Gulf oil spill:

1. The amount of people and equipment involved in the process was an unprecedented buildup for the U.S. Coast Guard to command and BP to oversee in an amazingly short period of time, a situation that was not covered in the existing BP response plan. This required a complexity-based system to mobilize and de-mobilize the vast resources necessary to combat the oil spill.

2. Traditional risk planning was abandoned in favor of on-the-spot response where management was able to respond dynamically, as the BP spill response plan was very poorly conceived and written and was unable to cope with the magnitude of the spill.
3. BP acknowledged the need to move the center of operations from a static base in Houston (as outlined in their response plan) to an on-the-spot management team empowered to make rapid decisions as new information came to light.
4. Local communications were identified as critical to the success of the spill cleanup.

In support of these findings, a review of the formal situation was done as well as a review of open-source literature, including newspaper articles, journal articles, trade journal articles, blogs, and other sources. All of these materials were examined to help understand the circumstances surrounding the incident. Considerable material has been written regarding the incident; however, much of the material focused on the sensational aspects of the event rather than on what could actually be learned from the event. In this regard, the authors of this book have chosen to maintain the focus on complexity, communication, and its application to the program.

Complexity to Support the Buildup of People, Materials, and Equipment

There is sufficient data to support the buildup of people, material, and equipment in a very short time. There was a smaller buildup at the beginning because of the inaccurate initial estimates of the magnitude of the spill. Early estimates were showing that the maximum amount of the spill was 5,000 barrels (National Commission, 2011), but this was shown to be a completely inaccurate forecast. This initial forecast hampered the response, because the BP spill response plan requires that an estimate be used as a first step (BP, 2010). These rigid elements hindered the initial attempts to contain the spill. The estimates were considered to be fundamentally accurate, which further caused delays in the deployment of additional equipment and material.

Because of the original estimate of 5,000 barrels per day (which was externally confirmed outside of BP), BP responded and deployed dispersant material to handle this size of a spill. When the initial deployment of dispersant was found to be inadequate, the only two possibilities were that the dispersant was defective or there was substantially more oil in the water that made the quantity of dispersant ineffective. Precious time was lost because the estimate was thought to be correct, and therefore the dispersant was being blamed as being defective. Once the dispersant was found to be in good working order, the only possible alternative was that the 5,000 barrel estimate was inaccurate.

At this juncture, BP should have considered both possible options instead of considering only the dispersant defect. The decision to look linearly at the problem rather than work on multiple possibilities shows how the limited risk plan hampered the initial attempts at spill containment. Early failures like this helped BP move from following the original risk plan to a more dynamic method of addressing the issue. The tragedy is that it took so long for BP to realize this deficiency and only later did its leadership accept the magnitude of the spill and move to respond appropriately.

"Complex systems almost always fail in complex ways" (National Commission, 2011, p. vii). The risk plan developed by BP was fundamentally flawed (even if one were to ignore the obvious glaring mistakes, such as information on how to handle sea lions and walruses in the spill plan, and the fact that the environmental consultant identified was dead years before the spill plan was submitted) in that the plan spends too much time discussing a linear approach to an oil spill (National Commission, 2011).

The plan includes pages of flowcharts on how and when to deploy dispersants for an oil spill. The deployment of dispersant is only considered if the oil is heading toward shore or colonies of sea birds. No other wildlife is considered as part of this process. If the flowchart allows for dispersants to be used, there is an internal requirement where approval must be sought in order to apply dispersants. This linear requirement seems to ignore that on-the-spot action might serve better than a hierarchical and linear response. Only after the spill was flowing out of control and national attention had been attained did BP start to move with more alacrity.

It is interesting to note that even after BP submitted paperwork as the responsible party on April 24, 2010, the mobilized response was quite small. On April 28, 2010, when the U.S. Coast Guard identified BP as the responsible party, the mobilization effort began in earnest. Within two days, BP moved from 500 people deployed (mostly for call centers to handle claims) to 2,000 people and 75 vessels deployed on April 30, 2010. It is clear that a shift occurred within BP management that the Gulf spill was no longer a simple matter of closing the well, applying dispersants, and deploying skimmers. The leak was larger than anyone had anticipated, and the process to clean up the spill and to close the well would take months. No simple risk plan flowchart would be able to cover a situation of this magnitude.

Abandonment of Traditional Hierarchical Communication

The order of magnitude of the spill and the escalation of the response are shown in Figure 16.1. This shows the rapid buildup of people and vessels involved in the cleanup efforts. In addition, BP deployed anywhere from 24 to 120 aircraft to assist with spotting spills in order to direct the vessels toward the oil.

It is important to reflect on Figure 16.1 to appreciate the magnitude of this situation. BP had a limited presence in the U.S. Gulf, mostly regarding the exploration of oil in the region, and because of the oil spill grew to be the single largest employer in the region, with up to 46,000 people involved with spill response. Not only did it have this large human resource, but BP also had to contract and charter over 5,000 vessels and over 100 aircraft at the peak of the incident (BP, 2011). Consider how large of a task this would be to be able to flex to a point that an organization with a limited presence would go to engaging 46,000 people all focused upon a single task—the containment of an oil spill.

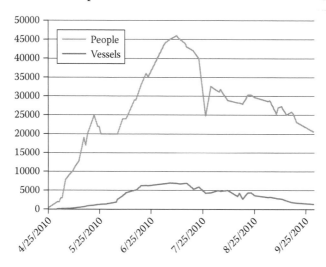

FIGURE 16.1
Rapid buildup of people and vessels involved in the cleanup efforts. (Curlee and Gordon, 2011. With permission from PMI.)

To understand the magnitude of the number of vessels deployed, consider that according to 2009 estimates, China had approximately 300 vessels in its navy and Australia had 51 vessels in the Royal Navy (Measuring the Chinese fleet, 2010; Royal Australian Navy, 2009). This means that BP had to move from no ability to manage a fleet of vessels to being able to mobilize, outfit, and organize a fleet of over 5,000 vessels. As the example of China and Australia shows, nations normally have considerably more time, tradition, and resources to organize a surface fleet, while BP had to find a way to command a fleet of over 5,000 vessels almost overnight.

Furthermore, consider the de-mobilization requirements necessary to address this change. Clearly, a complexity-based system needed to be developed in order to address these requirements. There is no doubt that a traditional system of hiring and chartering had to be abandoned in order to support such a huge buildup (Deepwater Horizon, 2010).

Very quickly the hierarchy of BP was overwhelmed by the multitude of requirements imposed by the national, state, and local governments, coupled with the negative press and ill feeling generated in the area. BP became vilified for the spill, and the company was seen as uncaring to the plight of those affected. To combat this public relations disaster, it required a complex and dynamic system to address all the communication challenges encountered. BP not only had to address the media, but they had to spearhead the capping of the spill and coordinate the environmental relief efforts associated with the spill.

Example of How Complexity Was Deployed to Replace Traditional Communication

BP employed complexity because they were forced to show results before the project team might be ready. This was seen with the media coverage of the Top Kill project. The project is the process of pumping a thick liquid into the leak, followed by cement. Top Kill was also unsuccessful with another spill, but the impatience of public opinion forced this component project to move forward. Any veteran program manager will agree that stakeholder impatience and haste can often create the necessity for non-sequential activities. A linear program manager might be paralyzed by this need to show results. In this case, the program manager at BP offered excuses that represented reality, but the need for action was seen as more important than the more likely solution, drilling a relief well.

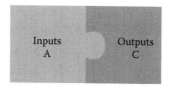

FIGURE 16.2

Inputs lead to outputs. (From Curlee, W. & Gordon, R. (2010). *Complexity theory and project management.* Hoboken, NJ: Wiley. With permission.)

Program managers can be pushed to resolve and handle issues out of the typical sequence in order to achieve certain milestones which are important at a higher level (Weaver, 2007). Political and media pressures exerted upon the company will flow to the program manager to achieve certain milestones faster. BP fell into this trap and was forced into a situation where inputs lead to outputs (results) (Figure 16.2).

BP eventually used a complexity-based strategy by starting a campaign to address the negative press. BP started advertising its organizational values to better explain how the situation would be addressed. It began leveraging its network in order to address this troubled project. There is no doubt this is a difficult decision for the organization, as it exposes it to potential short-term failure, but if the organization is expecting society to offer open and honest communication, then the organization must be guided by that same value (Duarte & Snyder, 2006). The result of not adhering to the value of open communication is far worse than any smaller issue which may arise from this type of problem. If one expects to use complexity to assist with these types of systemic and social challenges, one must be certain the organization, in this case BP, continues to model and share the values of the organization.

Local Deployment of an Incident Communication Center

To support this massive buildup and scale back, it was necessary for BP to establish an incident command post (ICP) in Houma, Louisiana. As the situation evolved, additional ICPs were established in order to manage the scale and uncertainty (complexity) of the spill. Over time, the response teams altered the established response structure to ensure support and authority flowed rapidly to local leadership and communities. This was critical as it allowed leveraging the local community in the cleanup efforts. A complex and comprehensive view of the location of the surface oil coupled with robust communication among leadership, airborne units, and responders on the water became the most effective way to deploy the necessary resources. As the spill continued, there was considerable concern regarding hurricane season approaching, and the local team needed to flex its long-term plan to address such issues as boom and skimmer supply and placement and inclement-weather operations (Deepwater Horizon, 2010).

Considering what needed to be done, it is not surprising that BP would give up centralized operations in Houston for a local system to support the local efforts. Since communication and location identification of deployed vessels were not easy to coordinate, it was better to have a local command post to address the local situation (Deepwater Horizon, 2010). In addition, allowing for a local command post with local support and resources was better accepted than having some distant operations dictate decisions. It is conceivable with ample communications and vessel location identification that a central location such as Houston might have been effective. The issue was that having a distant central authority gave the appearance of a lack of addressing the local issues at hand.

Consider that finding the spilled oil was difficult, and in turn deploying the right vessels to address the situation became equally challenging. Then contemplate balancing supply chain and human resources issues that come with any large-scale, unplanned deployment (BP, 2011; Deepwater Horizon, 2010). Trying to control such a monster at a great distance would be considerably difficult under the best of circumstances, and it would be completely impossible in any kind of unexpected circumstance. Departing from a linear and hierarchical system in favor of a local management structure empowered to make quick decisions became a key learning and necessary element in the future.

Figure 16.3 illustrates how the response escalated over time. The initial response hampered by poor information about the spill and the lackluster response by BP are clear by the lack of oil skimmed or burned. Figure 16.3 shows that clearly the stride of the response efforts came about 30 days after the initial sinking of the Deepwater Horizon. One can see the relationship between the change to a distributed, local management and the spill response effectiveness. As BP responded locally, the efforts became more focused and directed. A dispersed management team that was focused locally became the most effective management methodology to combat the spill.

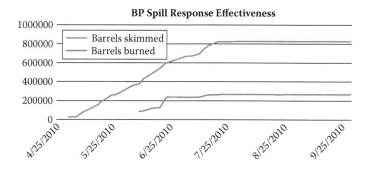

FIGURE 16.3

BP spill response effectiveness (as measured by barrels skimmed/burned). (Curlee and Gordon, 2011. With permission from PMI.)

The New Communication Model at BP

One of the crucial lessons learned was that in order to be effective in locating and cleaning up the oil, there had to be an effective manner with which to coordinate the vast amount of equipment and resources necessary to combat the spill. Since spill containment requires a variety of tasks (application of dispersant, booming, locating oil, moving equipment to the located oil), a local containment team needed to be present and in continuous communication. Also, the support fleet surrounding the spill site required coordination of drilling rigs and multipurpose support vessels, from fireboats to tugs, all operating in close proximity under hazardous conditions due to the presence of volatile hydrocarbons in the water.

Since all prior BP operations have involved only three or four ships in proximity to a platform, considerable planning had gone into any time BP entered into a situation that required simultaneous operations. Since vessels would have to operate within 40 feet of one another, the positioning of vessels was critical, and all movements were planned out well in advance.

In order to cap the Macondo spill, (the Deepwater Horizon oil spill) the ultimate operation required the simultaneous operation of up to 19 major vessels, each up to 825 feet in length, within a 1,650 foot radius of the wellhead, and up to another 50 vessels operating in a one mile radius. If this were not complicated enough, there would be times that vessels would be operating within 25 feet of one another. Given the complex nature of vessel operations, it was critical for vessels to continually understand each other's location as well as to carefully coordinate all of the movements and actions of these vessels (BP, 2011).

Since planned movements were not possible in advance, this complex operation required a new system that would allow the dynamic movement of vessels in a safe and expedient manner. One of the improvements that assisted in the simultaneous movement of equipment was the establishment of a rotating on-site branch director. The on-site director was responsible for the 24/7 operations and worked in coordination with the Houston-based team. Simultaneous operation also leverages the continuous use of storyboarding to allow team members to visualize the precise positioning and maneuvering of vessels. This process is not new and is commonly used in military operations where generals and admirals are given a better understanding of military operations by being able to visualize the movement of equipment.

A technology innovation that was identified as important to the simultaneous operation was the Automatic Identification Software (AIS), where an on-site transponder is placed on vessels to allow individuals involved to have real-time visualization, identification, tracking, and positioning of vessels on graphical displays (Deepwater Horizon, 2010). Many but not all of the vessels in the incident area were equipped with this relatively new technology. BP clearly identified that need for this type of equipment to be installed on all vessels operating in a situation like this, as this type of technology would support the relief efforts. This improvement in identification and tracking allows for operations to evolve without pre-planning, as those in command of the operations would be able to move about vessels operating in close proximity with a degree of certainty and safety. This allows for vessels operating in difficult conditions to work together with less risk. This is a more complexity-based solution where vessels are offering positioning information rather than the traditional and hierarchical method where the vessel would radio in its position and movements.

Risk Management, Communication, and Complexity

Program managers and component project managers must deal with risks and opportunities that will continually have to be planned throughout the life cycle of the program. In addition, the program manager will have to provide the team with the tools and confidence to react to the unexpected in order to be successful in changing circumstances (Pritchard, 2005; Weaver, 2007). The Gulf oil spill was a program of an order of magnitude not seen before. Although BP started with a very linear approach, it became clear that rigid restrictions were not in the best interest of the program. The Gulf spill became a program that quickly attained global significance along with a need to show rapid results. Complexity theory recognizes that such complex programs require an open system, such as a beehive or an anthill, that can react to the unexpected quickly and revert back to the norm or to a new norm. An open system like an anthill is more likely to survive long term, as the open system is fundamentally better at adapting to new circumstances.

What was seen was that the program needed to integrate the unexpected into the program and that the role of the leader is to provide guidance in order to have the program completed in the shortest possible time. The program manager has to balance risk management between the traditional risk management process and preparing the team to expect the unexpected. Program managers understand risk planning does not prevent the unknown unknown (PMI, 2013a).

As a result, BP has updated its risk management plan that now incorporates the following elements of complexity theory:

1. The implementation of robust, proven systems and tools for planning and implementing the management of large numbers of vessels at extremely close quarters, including storyboarding and a centralized, onsite control regime (Deepwater Horizon, 2010).

   This dynamic and continuous communication is designed to avoid a static hierarchical system of communication. Instead of relying upon a hierarchy and chain of command, a complexity-based network communication is necessary for success (Curlee & Gordon, 2010).

2. The deployment of AIS as an enabling technology for real-time visualization and management of offshore marine operations (Deepwater Horizon, 2010).

   Dynamic technology to allow real-time visualization and management of the offshore fleet is critical to success. This is another manifestation of a network communication where vessels are automatically communicating their position, speed, and location without any hierarchical or linear communication.

3. Demonstrated protocols for directing vessel traffic in the presence of flaring, even with the continuous incidence of VOCs and the need to ensure that levels were below the lower explosive limits (LELs), as well as new techniques for managing the presence of these hazards (Deepwater Horizon, 2010).

Command moves from being solely an order-driven organization to one that looks holistically at the efforts. Command can offer direction, not only for traffic, but also for hazard avoidance, as it can offer information that can assist vessels to avoid hazards in the area. This global perspective is required in a complexity-driven organization.

Conclusion

Complexity is everywhere, and more than ever complexity can assist business in achieving greater results with fewer resources. The more managers can apply complexity, the better they will be able to manage others. Complexity is about understanding the small in a way that it can be applied to the large (Curlee & Gordon, 2010). The BP spill has shown that programs will become larger and more complex, and program managers who learn to leverage complexity will be able to handle larger mega-programs (McKinnie, 2007). Project management as a whole needs to recognize that complexity theory is now a force in program management, even though the PMBOK Guide (PMI, 2013a) and Program Standard (PMI, 2013a) currently lack information about complexity theory. There are many competing leadership techniques currently available to program managers; however, none offer the same applicability and flexibility as complexity. The Gulf oil spill disaster has shown that even large organizations can learn to successfully apply complexity theory to a program. The more that organizations can learn from their mistakes, the more successful they will be in the changing global environment.

APPLYING FUTURE TECHNOLOGIES

To understand where complexity and program management will be in the future, one needs to consider where they have been in the past. Complexity has moved from being a fringe theory to one that is going mainstream in project management and program management circles. As people are looking for development to improve organizations, cutting-edge leaders are finding the principles of complexity can be applied to any complex undertaking. Program management is being understood for both commercial undertakings and for the military.

Advanced program managers understand that skills and abilities beyond linear thinking are required to make a dispersed program successful. Linear control is no longer possible for large complex programs, so a leader must learn to be flexible and to harness the available technology to communicate through different time zones and over great distances. Technology has certainly changed, but more importantly the understanding of social media, online tools, and other technologies will shape program management and complexity in the future. It is no longer possible to manage a program with just face-to-face contact or a co-located group. Programs will have stakeholders, suppliers, and other participants that will be based at various locations not always at arm's reach of the program manager.

There is still much knowledge to be discovered with technology and complexity (Strickland, 2013). For example, the use of social media has shown how a single individual can impact thousands to millions of people as blogs, pictures, and videos go viral. Who would have expected that a dancing baby or Chum-Ninja would have millions of hits by casual Internet users looking for entertainment? Even with modern science, we may have theories on how the mind works and how ideas are formed, but we still have no answers as to why the mind can drive individuals to click a mouse to what we would normally would perceive to be silly, a dancing baby.

What we see now are program managers trying to harness this same technology to improve programs. Just as a program manager can work with a solid reclaiming, recycling, and re-using plan to save money with technology, different technology needs to be utilized to improve leadership and communication. In the end, the success of a program is based upon its people.

Good technology can assist with these goals, but if good people are not in charge, the technology will be squandered. A high-tech mobile phone with the capacity to use a wide variety of applications has the potential to organize and keep people in touch; however, if one does not use the applications, then it is still just a phone.

Similarly, Wal-Mart understood early on the potential of radio frequency identification (RFID) equipment and integrated it early into its supply chain process by requiring the use of this technology by its suppliers. This early adaptation forced suppliers to move with technology as well as it forced Wal-Mart to find the best ways to leverage this technology. This symbiotic relationship made for the technology to multiply in usefulness rather than lay dormant on the sidelines. Making new technology part of the process is one way to improve the process. The complexity to this is that Wal-Mart saw the potential but did not know exactly how it could be used. Once the technology was there, it was able to be used in multiple ways previously not conceived. Had the technology not been there and available, then it would not have evolved, but since it was required, it made sense for everyone to harness the technology in manners that made the entire process more efficient. In this same manner, program managers and complexity will evolve in the future.

Once program managers learn to use complexity and deploy it in their programs, others will learn to use the theory more strategically and it will be used further. New technology based upon social media will be utilized to support complex use, and it will continue to grow by leaps, rather than being pushed by certain individuals.

Technology will be leveraged by the younger generation in ways that the older generation could never have imagined. One must learn to embrace this change in order to be ready for the future. If one does not make the change, then one can expect that the competition will make the change and surpass the individual. The most dangerous strategy is to wait. Waiting means letting others take the strategic initiative and letting them dictate the new terms of engagement. If one is not willing to change and change quickly, then one will always be giving up the strategic initiative to others and they will be forced to try to catch up.

The more in depth one examines the complexity and program management, the more one needs to accept new technology, new ideas, and most of all the new leaders who will be willing to take the risks to meet the future. Tomorrow's leaders will communicate faster, be better understood, and be followed by millions. Consider how texting shorthand evolved over time without being taught in any school. For those who remember, shorthand was once a secretarial skill that was taught in schools, to quickly put spoken words to paper. Now millions of people worldwide understand text shorthand to a degree that it transcends language, nations, and cultures. That achievement cannot be explained without complexity.

ADVANCING COMPLEXITY

Complexity offers new solutions; however, it can also offer new challenges. Figure 16.4 shows how *forward* can be a matter of perspective more than an absolute.

Forward Can be a Matter of Perspective

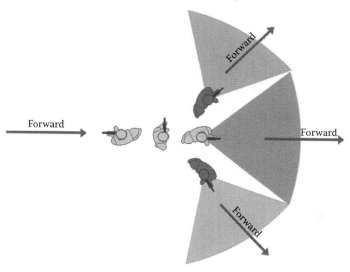

FIGURE 16.4
Which way is forward?

Keeping in mind that leadership needs to always communicate the absolute direction, review the advancing complexity tool presented in Figure 16.5. This can be applied to an existing program or to a potential program. One should also consider asking other stakeholders on the component projects to review and provide feedback on the tool to the project manager and the program manager.

Name: Date:

ADVANCING COMPLEXITY IN PROGRAM MANAGEMENT

Instructions	Expect to spend about 15–30 minutes with this tool. Try not go into long details but make sure that the idea and the plan are easily understood by others.

STEP 1

Programs are inherently linear. Make a list of three ways to make a program less linear

STEP 2

Create a plan to implement item one into the program to advance complexity. Make sure that the plan has a goal and a deadline.

STEP 3

Create a plan to implement item two into the program to advance complexity. Make sure that the plan has a goal and a deadline.

STEP 4

Create a plan to implement item three into the program to advance complexity. Make sure that the plan has a goal and a deadline.

STEP 5

Advertise the plan to others so that others can assist with implementing the changes.

FIGURE 16.5
Advancing complexity tool.

17

Course Materials #5

PHOTOGRAPH 17.0
Castle Industries—Building the strongest vessels for the future.

CLASSROOM MATERIALS

Case Study #5—The Future of Program Management and Complexity

You are a program manager for a large established company called Castle Industries that does a lot of military subcontracting. Your company has been engaged to design the next generation of supply vessels to support military operations in other nations. However, this vessel needs to be designed with two important differentiations from a normal warship. First, it needs to be designed so that it can convert from a supply vessel to a troop transport to a medical vessel to a vessel to support relief operations. Second, the vessel needs to be designed in a manner that it can be built in modules in different shipyards and then assembled in one location. Due to the complexity of the design and engineering of this vessel, the United States has commissioned a five-year design program to develop the vessel that accommodates these needs.

As the program manager for this company, you have been tasked with building a team of ten professionals from any part of the company to handle each of the following aspects of the engineering program (Figure 17.1):

Lead engineer
Contract manager
Financial manager
Creative design
Client support

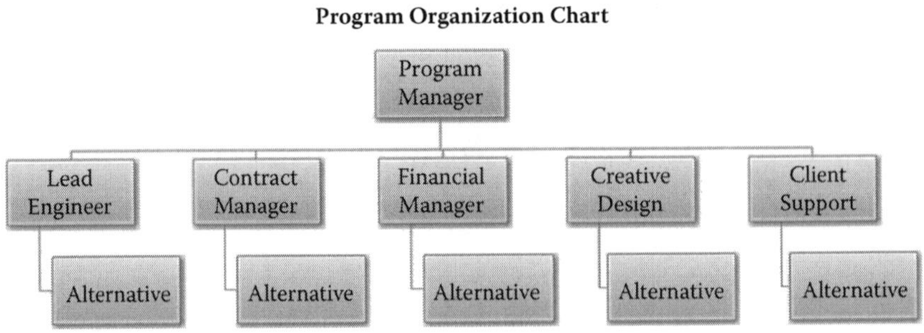

Program Organization Chart

FIGURE 17.1
Organizational chart.

You will need to select a primary and a secondary for each role in the event that some of the key people leave the company or are pulled from the project. Since this design will span five years, your experience tells you that you need to have a primary and a secondary to make sure that there is continuity throughout the five years of the program. This program is considered high profile, and so you need to have someone in client support to assist with making sure that the client is kept appraised of progress. Although this is your permanent assignment, you know that you will also be tasked with other requirements by the company.

Company Details

The company is called Superior Ship Design, and your company specializes in multipurpose military vessels. The company has 15 years of service to the military and other shipping companies. The primary business has been with the military, and the company does work in the United States and internationally.

Superior Ship Design is based in the Houston area, but some of the employees are located in other areas as the company has a tradition of hiring home-based employees to gather the best talent available without relocating them. This strategy has proven very successful as they can keep costs down while keeping employee satisfaction up. The company has a number of people available to work with you; all of them will ultimately report to you. You are based in Houston and the company has given you this role as long as you remain in Houston, as the owners want to be in continual contact with you to ensure that this program is successful.

Summary of Employees

Jason Anders

Jason has been with the company for 10 years and has been the project head of several successful projects. He is well liked and well regarded in the organization. He is considered organized and driven. He has a strong customer focus, and he is known to be able to handle the toughest clients. He can have a short temper with others internally, but he is known for getting the tough jobs done. He does not get along with Cindy Jolive and is vocal about not wanting to work on the same project as her. He is currently located in Seattle, Washington.

Bob Bennancourt

Bob has five years with the company and is considered a great engineer. He is known for finding good solutions to complex problems. Most people can work with him, and he can be a good mentor to new engineers. He has been known to overdesign ships and sometimes needs to be monitored closely. He has worked for NASA and often refers to that experience. He is based in the Houston area.

Carl Clear

Carl has been with the company for 10 years and is considered a top engineer. He has been on many successful projects, most of them related to design. He is respected by most people in the company and is a great team player. He can resist change to his design for cost reasons, but he is known to be good for completing the job on time. He is based in the Dallas area.

Dom Franks

Dom has nine years with the company and has been involved with several successful projects. He is considered a good financial controller and can garner stakeholder support. He is not afraid to spend money to save it in the long run. He is a hard worker and is dedicated, but he can get distracted with pet projects. He works well on his own and sometimes resists being part of a team. He is based in Orlando, Florida.

Tom Gardener

Tom has only been with the company for two years and has little experience with leading a full program. He has done some contract management work with another company for five years and is considered a good negotiator. He has clashed with Carl Clear in the past over some engineer costs and vendor selection. Tom is based in Houston.

Cindy Jolive

Cindy has 10 years with the company and has been a successful project manager for a number of larger projects. She has five years of experience as a contract manager with another company, and although she has done well for the company she can be very direct with people. Her demeanor has put off some clients in the past, but she always gets results. Cindy and Jason Anders do not get along and will actively avoid one another. She is based in Houston.

Tammy Mason

Tammy has been with the company for four years and is mostly involved with creative design. Her designs have won several awards in the past. She is considered to be well organized, and most people enjoy working with her, except Carl Clear. She can be very social, and she has many positive reviews from clients. Tammy is currently based in New York City.

Ron Nieber

Ron has eight years of contract management experience and has been with the company for 10 years. He started in finance but quickly worked toward contract management. Ron has a good reputation within the company and can work with everyone. Ron likes to work with Carl Clear, and the two of them have been on some successful programs together. Ron is based in Houston.

Todd Morison

Todd has 15 years of experience with programs and has led a few successful programs. He was a contract manager during his last program, and in addition to this role he did a good job filling in for the program manager when he was unavailable. Todd is very easygoing but can come down hard on poor performers. Todd can accept things running a little late, but he cannot accept people who do not put in long hours. He feels that people who don't put in the long hours are lazy, and this has put off several people in the past. He is based in Atlanta.

Paul Paulson

Paul has four years of experience with the company as an engineer. He has worked with a couple of other competitors prior to joining the company, and he has some diverse experience. He is considered creative and can work with everyone. He can be financially focused if directed that way but can overdesign if left unmonitored. Pat is a relatively untested leader, but it is felt that he could do well given the right team. Paul is based in Houston.

Sam Roberson

Sam has seven years of client services experience with the company and another five years with another company. He is viewed as a good client services support person and has worked on several large programs. Sam is good with smoothing things over with clients, but he tends to overcommunicate with the customer. He is a very motivated individual, and he thinks highly of Paul Paulson. Sam is based in Dallas.

Rod Smith

Rod has 12 years of experience with the company in finance. In addition to his financial background he is focused upon safety. He has been involved with a few successful programs and enjoys working with Carl Clear. He is a driven employee who is seen as getting results. He is based in Houston.

Larry Tandle

Larry has 10 years in finance with the company and is considered a financial wizard by many people in the company. He is known to be better than anyone else at making the numbers look good to the client. He has worked hard to keep his reputation solid and is well liked within the organization. He is based in San Diego but has spoken about moving to Houston in the near future.

Julie Tyler

Julie has four years of design experience with the company and has also served as a financial manager in the past. She is very focused on the financial health of a program but also enjoys design. She has done some solid design work, but most of her work recently has been in finance. She prefers not to work with Rod Smith as they did not work well together on the last program. Julie is based in Erie, Pennsylvania.

Linda Vera

Linda is a new employee and was doing design for a competitor for 10 years before moving to the company. She is considered strong willed and determined, and people feel that she will contribute great new designs in the future. She has not been part of a team yet in the company, but most people like her. She is based in Houston and has expressed a keen interest in being part of this program.

Thomas Walters

Thomas has 10 years with the company and is very financially driven. He is good with numbers and is seen as an asset due to his ability to handle accounting problems. He is a professional finance person but has been known to get involved with contracts. He is shy in meetings but is good at working behind the scenes to get things done. He is not always good with clients because they see his shyness as an inability to communicate. He does not like to work with Bob Bennancourt because he feels that Bob brags more than he works. Thomas is based in Houston.

Assignment

1. Given the program, identify the people and the succession plan for the following roles:
   a. Lead engineer
   b. Contract manager
   c. Financial manager
   d. Creative design
   e. Client support
2. Defend your selections and offer clear points as to why certain people were chosen over others.
3. Explain the leadership plan that you will use to lead this team.
4. Explain the communication plan that you will use to communicate with the team and the client.
5. Detail how you will manage, control, and maintain the virtual organization if your entire team is not based in Houston.
6. What elements of complexity would you deploy to assist the success of the program team?

PHOTOGRAPH 17.1
Putting it all together.

SECTION QUIZ

Section 1

Multiple Choice

1. British Petroleum changed their communication during the Gulf oil spill because:
   a. The president insisted that they cover the costs of the spill.
   b. The Coast Guard required the change.
   c. The amount of people was unprecedented.
   d. Technology could not handle it in the prescribed manner.

2. Complexity theory recommends communication become:
   a. Centralized
   b. Controlled
   c. Local
   d. None of the above

3. Complexity theory offers a change in perspective in risk management in a manner that recommends:
   a. Planning tools and proven systems
   b. Technology
   c. Off-site management
   d. All of the above

Section 2

True/False

1. British Petroleum kept and maintained a proven communication system.
   a. True
   b. False

2. Program managers can learn from the mistakes of prior programs.
   a. True
   b. False

3. The latest technology is always best for quality communications.
   a. True
   b. False

4. Complexity theory recommends a holistic approach.
   a. True
   b. False

5. A complexity-based project should always maintain a static center of operations.
   a. True
   b. False

Section 3

Answer Key

Section 1

   1. C
   2. C
   3. A

Section 2

   1. B
   2. A
   3. B
   4. A
   5. B

DISCUSSION QUESTIONS

Short Discussion Questions

Can complexity theory and program management work together to achieve greater results?

Will complexity theory be applied more or less to program management in the future?

PHOTOGRAPH 17.2

The future of complexity and program management is about making synchronous and positive programs in a rapidly evolving asynchronous world. Working together toward a common goal can yield impressive results.

References

Anderson, E., D. Doyle, P. Friedlander, D. Schroeder, and T. Seymour. (1998). Telecommuting primer. *Infotech Update, 7*(7), 1.

Andraski, J. C. (1998). Leadership and the realization of supply chain collaboration. *Journal of Business Logistics*.

Badaracco, J. L., Jr. (1998). The discipline of building character. In *Harvard business review on leadership* (pp. 89–113). Boston: Harvard Business School Press.

Bass, B. (Ed.). (1990). *Bass & Stogdill's handbook of leadership* (3rd ed.). New York: The Free Press.

Bass, B. M., & Riggio, R. E. (2006). *Transformational leadership* (2nd ed.). Mahwah, NJ: Lawrence Erlbaum.

Bennis, W. (1994). *On becoming a leader*. New York: Addison-Wesley.

Bennis, W., & Goldsmith, J. (1997). *Learning to lead* (Updated ed.). Reading, MA: Perseus.

Bennis, W., & Nanus, B. (1997). *Leaders: Strategies for taking charge* (2nd ed.). New York: HarperBusiness.

Blanchard, D. (2012, February). Going in reverse can be the right direction: Returns management can offer significant cost savings for manufacturers. *Industry Week*. Available at http://www.industryweek.com/articles/going_in_reverse_can_be_the_right_direction_26594.aspx?SectionID=2.

Boudreau, M., Loch, K., Robey, D., & Straud, D. (1998). Going global: Using information technology to advance the competitiveness of the virtual transnational organization. *Academy of Management Executive, 12*(4), 120.

BP oil spill response plan—Gulf of Mexico. (2010). Retrieved January 29, 2011, from http://info.publicintelligence.net/BPGoMspillresponseplan.pdf.

Brier, B. (2002). The other pyramids. *Archaeology, 55*(5), 54.

Brown, K. L. (2000, September). Analyzing the role of the project consultant: Cultural change implementation. *Project Management Journal, 31*(3), 52–55.

Byatt, G. (2013). Demystifying stakeholder management: The science and the art. In G. Levin (Ed.), *Program management: A life cycle approach* (pp. 129–142). Boca Raton, FL: Auerbach.

Byrne, D. (1998). *Complexity theory and the social sciences: An introduction*. New York: Routledge.

Cascio, W. (2000). Managing a virtual workplace. *The Academy of Management Executive, 14*(3), 81–90.

Checkland, P. (1999). *Systems thinking, systems practice*. West Sussex, England: Wiley.

Chemers, M. (1995). In T. Wren (Ed.), *The leader's companion: Insights on leadership through the ages* (pp. 83–99). New York: The Free Press.

Chui, M., Manyika, J., Bughin, J., Dobbs, R., Roxburgh, C., Sarrazin, H., Sands, G., & Westergren, M. (2012). *The social economy: Unlocking value and productivity through social technologies*. New York: McKinsey & Company.

Cohen, E., & Tichy, N. (1999, September). Operation leadership. *Fast Company*, Boston, MA: Fast Company Media Group.

Collins, J. C., & Porras, J. I. (1997). *Built to last: Successful habits of visionary companies*. New York: HarperCollins.

Cooke-Davies, T., Cicmil, S., Crawford, L., & Richardson, K. (2007). We're not in Kansas anymore, Toto: Mapping the strange landscape of complexity theory, and its relationship to project management. *Project Management Journal, 38*(2), 50–61.

Curlee, W., & Gordon, R. (2010). *Complexity theory and project management.* Hoboken, NJ: Wiley.

Curlee, W., & Gordon, R. (2011). Risk management through the lens of complexity. 2011 PMI EMEA Global Congress North America Proceedings, Dallas, TX.

Dahlgaard, S. P., Dahlgaard, J. J., & Edgeman, R. L. (1998, July). Core values: The precondition for business excellence. *Total Quality Management, 9*(45), 51–55.

Dani, S., Burns, N., Backhouse, C., & Kochhar, A. (2006). The implications of organizational culture and trust in the working of virtual teams, *Proceedings of the Institution of Mechanical Engineers, Part B, Engineering Manufacture (Professional Engineering Publishing), 220*, 951–959.

Deepwater Horizon containment and response: Harnessing capabilities and lessons learned. (2010, September). Retrieved January 30, 2011, from http://www.bp.com/liveassets/bp_internet/globalbp/globalbp_uk_english/incident_response/STAGING/local_assets/downloads_pdfs/Deepwater_Horizon_Containment_Response.pdf.

Department of the Army. (2006). Field Manual 6-22. Army Leadership; Competent, Confident, and Agile. Secretary of the Army.

Dessler, G. (2001). Management: Leading people and organizations in the 21st century. (2nd ed.). Upper Saddle River, NJ: Prentice Hall.

Drucker, P. (1996). *The effective executive.* New York: HarperCollins.

Duarte, D., & Snyder, N. (1999). *Mastering virtual teams* (2nd ed.). San Francisco: Jossey-Bass.

Duarte, D., & Snyder, N. (2006). *Mastering virtual teams* (3rd ed.). San Francisco: Jossey-Bass.

Elkins, T. (2000). Virtual teams. *IIE Solutions, 32*, 26–31. EBSCOhost (Masterfile Premier).

Gordon, M. (1965). Cramming more components into integrated circuits. *Electronics, 38*.

Gordon, R. (2011). *Reverse logistics management.* Hoboken, NJ: Wiley.

Gordon, R., & Curlee, W. (2011). *The virtual project management office: Best practices, proven methods.* Vienna, VA: Management Concepts.

Grier, D. (2013). Technologies of the future: 5 Trends to watch for 2013. *Forbes Online.* http://www.forbes.com/sites/ericsavitz/2012/12/07/technologies-of-the-future-5-trends-to-watch-for-2013/

Handy, C. (1995). Trust and the virtual organization. *Harvard Business Review, 73*(3), 40.

Hass, K. (2009). *Managing complex projects: A new model.* Vienna, VA: Management Concepts.

Heifetz, R. A., & Laurie, D. L. (1998). *The work of leadership.* In Harvard Business School Press (Ed.), *Harvard business review on leadership* (pp. 171–197). Boston: Harvard Business School Press.

Hemmingway, J. (1998). *Disney Magic: The launching of a dream.* New York: Hyperion Press.

Hemsath, D. (1998, January/February). Finding the word on leadership. *Journal for Quality and Participation, 21*(1), 50–51.

Hillman, J. (1996). *The soul's code: In search of character and calling.* New York: Warner.

Holmes, R. (2012). The can't-miss social media trends for 2013. FastCompany. http://www.fastcompany.com/3003473/cant-miss-social-media-trends-2013

Hyslop, J. (1984). *The Inca road system.* Orlando, FL: Academic Press.

Jaafari, A. (2003). Project management in the age of complexity and change. *Project Management Journal, 34*(4), 47–57.

Jacques, R. (1996). *Manufacturing the employee: Management knowledge from the 19th to 21st century.* London: Sage.

Jarvenpaa, S., & Leidner, D. (1999). Communication and trust in global virtual teams. *Organization Science, 10*(6), 791.

Johnson, D., & Johnson, F. (2000). *Joining together: Group theory and group skills.* Needham Heights, MA: Allyn & Bacon.

Joy-Matthews, J., & Gladstone, B. (2000). Extending the group: A strategy for virtual team formation. *Industrial and Commercial Training, 32,* 24–29.

Karl, K. (1999). Mastering virtual teams book review. *The Academy of Management Executive, 13,* 118–119.

Kelley, E. (2001). Keys to effective virtual global teams. *Academy of Management Executives, 15*(2), 132.

Kent-Drury, R. (2000, February). Bridging boundaries, negotiating differences: The nature of leadership in cross-functional proposal-writing groups. *Technical Communication, 41*(1), 90–98.

Kirkpatrick, S. A., & Locke, E. A. (1995). Leadership: Do traits matter? In T. Wren (Ed.), *The leader's companion: Insights on leadership through the ages* (pp. 133–143). New York: Free Press.

Krajewski, L., & Ritzman, L. (2001). *Operations management, strategy and analysis* (5th ed.). Reading, MA: Addison-Wesley.

Levin, G. (Ed.). (2012). *Program management: A life cycle approach.* Boca Raton, FL: CRC Press.

Li, X., & Olorunniwo, F. (2008). An exploration of reverse logistics practices in three companies. *Supply Chain Management: An International Journal* (13.5), 381–386.

Lichtenstein, B., Uhl-Bien, M., Marion, R., Seers, A., Orton, J., & Schreiber, C. (2006). Complexity leadership theory: An interactive perspective on leading in complex adaptive systems. *E.CO (8),* 4. 2–12

Lipnack, J., & Stamps, J. (1999). Virtual teams: The new way to work. *Strategy & Leadership, 27*(1), 14.

Lipnack, J., & Stamps, J. (2000). *Virtual Teams: People working across boundaries with technology* (2nd ed.). New York: Wiley.

Lorenz, E. (1972). Predictability: Does the flap of a butterfly's wings in Brazil set off a tornado in Texas? American Association for the Advancement of Science, 139th Meeting.

MacPhail, J. (2007). Virtual teams: Secrets of a successful long-distance research relationship, a Canadian perspective. *Annals of Family Medicine, 5*(6), 568–569.

Marcinko, R., & Weisman, J. (1997). *Leadership secrets of the rogue warrior: A commandos guide to success.* New York: Simon & Schuster.

Marx, K., & Engels, F. (1978). Manifesto of the communist party. In R. C. Tucker (Ed.), *The Marx-Engels reader* (2nd ed.), (pp. 468–500). New York: W. W. Norton.

Maznevski, M., & Chudoba, K. (2000). Bridging space over time: Global virtual team dynamics and effectiveness. *Organization Science: A Journal of the Institute of Management Sciences, 11*(5), 473.

McKinnie, R. (2007). The application of complexity theory to the field of project management. (UMI doctorate dissertation, No. 3283983).

Measuring the Chinese fleet. (2010, January). Retrieved January 29, 2011, from http://www.strategypage.com/htmw/htsurf/articles/20100121.aspx.

National Air Traffic Controller Association. (2012). Sequestration: The effects on aviation and everyday travel; how sequestration will affect the flying public and the U.S. economy.

National Commission on the BP deepwater horizon oil spill and offshore drilling. (2011, January). *Deep Water: The Gulf disaster and the future of offshore drilling. The Report to the President.* Retrieved January 30, 2011, from https://s3.amazonaws.com/pdf_final/DEEPWATER_ReporttothePresident_FINAL.pdf.

Nohria, N., & Berkley, J. D. (1998). Whatever happened to the take-charge manager? In *Harvard business review on leadership* (pp. 199–222). Boston: Harvard Business School Press.

O'Connor, C. (2000, August). Building the virtual team. *Accountancy Ireland, 32,* 20–21.

Overman, E., & Loraine, D. (1994). Information for control: Another management proverb. *Public Administration Review, 54*(2), 193–196.

Platt, L. (1999, September/October). Virtual teaming: Where is everyone? *Journal for Quality & Participation, 22,* 41–43.

Pritchard, C. (2005). *Risk management: Concept and guidance* (3rd ed.). Arlington, VA: ESI International.

Project Management Institute (Ed.). (2013a). *A guide to the project management body of knowledge.* Newtown Square, PA: PMI Publishing Division.

Project Management Institute (Ed.). (2013b). *The program management standard.* Newtown Square, PA: PMI.

Reinsch, L. (1999). Selected communications variables and telecommuting workers. *Journal of Business Communications, 36*(3), 247.

Roberts, K., Kossek, E., & Ozeki, C. (1998). Managing the global workforce: Challenges and strategies. *Academy of Management Executive, 12*(4), 93.

Roebuck, D. (2001). *Improving business communication skills* (3rd ed.). Upper Saddle River, NJ: Prentice Hall.

Rogers, D., & Tibbens-Lembke, R. (1998). *Going backwards: Reverse logistics trends and practices.* Reverse Logistics Executive Council. Available at http://www.rlec.org/reverse.pdf.

Royal Australian Navy. (2009). In *Wikipedia, the free encyclopedia.* Retrieved January 29, 2011, from http://en.wikipedia.org/wiki/Royal_Australian_Navy.

Samoilenko, S. (2008). Information systems fitness and risk in IS development: Insights and implications from chaos and complex systems theory. *Information Systems Front, 10,* 281–292.

Saynisch, M. (2010). Beyond frontiers of traditional project management: An approach to evolutionary, self-organizational, principles and the complexity theory—Results of the research. *Project Management Journal, (41)* 2, 21–37. DOI: 10.1002.

Schein, E. H. (2004). *Organizational culture and leadership* (3rd ed.). San Francisco: Jossey-Bass.

Shafritz, J., & Ott, J. (1996). *Classics of organization theory* (4th ed.). Ft. Worth, TX: Harcourt Brace College.

Simpson, J. (1970). The Polaris executive. *Public Administration, 48*(4), 379.

Stock, J., & Mulki, J. (2009). Product returns processing: An examination of practices of manufacturers, wholesalers/distributors and retailers. *Journal of Business Logistics, 30*(1), 33–62. ABI/INFORM Global.

Strickland, J. (2013). What is the future of communication? How stuff works. http://electronics.howstuffworks.com/everyday-tech/future-of-communication.htm

Tavcar, J., Zavbi, R., Verlinden, J., & Duhovnik, J. (2005). Skills for effective communication and work in global product development teams. *Journal of Engineering Design, 16*(6), 557–576.

Thomas, J. (2000). Making sense of project management: Contingency and sensemaking in transitory organization. Proquest Digital dissertation.

Townsend, A., & DeMarie, S. (1998). Keys to effective virtual global teams. *Academy of Management Executive, 12*(3), 17–30.

Uhl-Bien, M., & Marion, R. (2009). Complexity leadership in bureaucratic forms of organizing: A meso model. *Leadership Quarterly,* 20, 631–650.

U.S. Army and Marine Corps (Ed.). (2007). Counterinsurgency Field Manual. Chicago, IL: University of Chicago Press.

Useem, M., & Harder, J. (Winter 2000). Leading laterally in company outsourcing. *Sloan Management Review 41*(2), 25–36.

Waltuck, B. (2012). Thriving in a changing world: Preparing today for an uncertain tomorrow. *Journal for Quality & Participation, 35*(3), 13–16.

Weaver, P. (2007). A simple view of complexity in project management. PMOZ Conference Keynote address.

Weiss, W. H. (1999, January). Leadership. *Supervision, 64*(11), 17–20.

Wheatley, M. (1999). *Leadership and the new science: Discovering order in a chaotic world.* San Francisco: Berrett-Koehler.

Index

Printed and bound by CPI Group (UK) Ltd, Croydon, CR0 4YY

23/10/2024

01777673-0005